The Business of Global Governance

Global Environmental Accord: Strategies for Sustainability and Institutional Innovation
Nazli Choucri, series editor

Nazli Choucri, editor, *Global Accord: Environmental Challenges and International Responses*

Peter M. Haas, Robert O. Keohane, and Marc A. Levy, editors, *Institutions for the Earth: Sources of Effective International Environmental Protection*

Ronald B. Mitchell, *Intentional Oil Pollution at Sea: Environmental Policy and Treaty Compliance*

Robert O. Keohane and Marc A. Levy, editors, *Institutions for Environmental Aid: Pitfalls and Promise*

Oran R. Young, editor, *Global Governance: Drawing Insights from the Environmental Experience*

Jonathan A. Fox and L. David Brown, editors, *The Struggle for Accountability: The World Bank, NGOS, and Grassroots Movements*

David G. Victor, Kal Raustiala, and Eugene B. Skolnikoff, editors, *The Implementation and Effectiveness of International Environmental Commitments: Theory and Practice*

Mostafa K. Tolba, with Iwona Rummel-Bulska, *Global Environmental Diplomacy: Negotiating Environmental Agreements for the World, 1973–1992*

Karen T. Litfin, editor, *The Greening of Sovereignty in World Politics*

Edith Brown Weiss and Harold K. Jacobson, editors, *Engaging Countries: Strengthening Compliance with International Environmental Accords*

Oran R. Young, editor, *The Effectiveness of International Environmental Regimes: Causal Connections and Behavioral Mechanisms*

Ronie Garcia-Johnson, *Exporting Environmentalism: U.S. Multinational Chemical Corporations in Brazil and Mexico*

Lasse Ringius, *Radioactive Waste Disposal at Sea: Public Ideas, Transnational Policy Entrepreneurs, and Environmental Regimes*

Robert G. Darst, *Smokestack Diplomacy: Cooperation and Conflict in East-West Environmental Politics*

Urs Luterbacher and Detlef F. Sprinz, editors, *International Relations and Global Climate Change*

Edward L. Miles, Arild Underdal, Steinar Andresen, Jørgen Wettestad, Jon Birger Skjærseth, and Elaine M. Carlin, *Environmental Regime Effectiveness: Confronting Theory with Evidence*

Erika Weinthal, *State Making and Environmental Cooperation: Linking Domestic and International Politics in Central Asia*

Corey L. Lofdahl, *Environmental Impacts of Globalization and Trade: A Systems Study*

Oran R. Young, *The Institutional Dimensions of Environmental Change: Fit, Interplay, and Scale*

Tamar L. Gutner, *Banking on the Environment: Multilateral Development Banks and Their Environmental Performance in Central and Eastern Europe*

Liliana B. Andonova, *Transnational Politics of the Environment: The European Union and Environmental Policy in Central and Eastern Europe*

David L. Levy and Peter J. Newell, editors, *The Business of Global Environmental Governance*

The Business of Global Environmental Governance

Edited by David L. Levy and Peter J. Newell

The MIT Press
Cambridge, Massachusetts
London, England

This book was set in Sabon by SNP Best-set Typesetter Ltd., Hong Kong. Printed on recycled paper and bound in the United States of America.

Library of Congress Cataloging-in-Publication Data

The business of global environmental governance / edited by David L. Levy and Peter J. Newell.
    p. cm.—(Global environmental accords)
Includes bibliographical references and index.
ISBN 0-262-12270-7 (hc. : alk. paper)—ISBN 0-262-62188-6 (pbk. : alk. paper)
        1. Environmental policy—Economic aspects. 2. Environmental policy—International cooperation. 3. Globalization—Environmental aspects. I. Levy, David L. II. Newell, Peter (Peter John) III. Series.

HC79.E5B8664 2005
333.7—dc22

2004049886

10 9 8 7 6 5 4 3 2 1

Contents

Series Foreword

A new recognition of profound interconnections between social and natural systems is challenging conventional constructs and the policy predispositions informed by them. Our current intellectual challenge is to develop the analytical and theoretical underpinnings of an understanding of the relationship between the social and the natural systems. Our policy challenge is to identify and implement effective decision-making approaches to managing the global environment.

The series Global Environmental Accord: Strategies for Sustainability and Institutional Innovation adopts an integrated perspective on national, international, cross-border, and cross-jurisdictional problems, priorities, and purposes. It examines the sources and the consequences of social transactions as these relate to environmental conditions and concerns. Our goal is to make a contribution to both intellectual and policy endeavors.

Nazli Choucri

Acknowledgments

This book was born out of a conversation between the co-editors at a 1997 conference at Warwick University on Non-State Actors and Authority in the Global System. We both happened to be studying corporate strategies for engaging with the climate change issue, and the impact of these strategies on the emerging regime. The idea for the book grew from the realization that our personal conversation could be developed into a fruitful interaction between our two disciplines, international relations and international political economy on one side, and management and organization theory on the other. We shared a strong sense that the role of business and industry in global environmental politics was a critical yet currently underdeveloped area of study.

Inevitably, the book has taken a lot longer to complete than either of us intended but we are delighted it has finally come to fruition. We hope that it will stimulate future thinking and conversations. We appreciate the hard work and patience of all our contributors in awaiting the final product and for putting up with endless demands for revisions. We would also like to thank Clay Morgan as commissioning editor for his guidance and support.

Peter Newell would like to thank colleagues who have helped to develop his thinking on questions of corporate accountability and regulation in relation to environment and development for a number of years now. These include David Levy, Matthew Paterson, John Gaventa, Dominic Glover, Ian Scoones, and Halina Ward. David, in particular, deserves special mention as an inspiring collaborator and committed colleague. I would also like to acknowledge the invaluable administrative support of Oliver Burch in the final stages of the book, and the love and friendship provided throughout by Bridget Allan, Rahul Moodgal, and my parents, Helen and Brian Newell. I dedicate this book to the memory of my late grandmothers, Vera Diehl and Rita Ferguson, who have always been an inspiration to me.

David Levy would like to thank colleagues who have collaborated on various papers related to multinational corporations and global environmental governance, and who have thereby participated in shaping and sharpening the conceptual framework. These include Daniel Egan, Ans Kolk, Peter Newell, Aseem Prakash, and Sandra Rothenberg. I would also like to thank colleagues at institutions that have funded the studies of the oil and automobile industries, which form the core empirical basis of my work in recent years. These include David Marks and Joanne Kauffman at the Consortium on Environmental Challenges, Massachusetts Institute of Technology, Bill Clark and the faculty and fellows of the Global Environmental Assessment Project, Kennedy School of Government, Harvard University, and Frans van der Woerd, Free University of Amsterdam. I particularly want to acknowledge Peter Newell's role in fully sharing the burden of editing this volume and helping to pull us through moments when I despaired that this volume would never reach the printers. I want to thank Gail Dines for her love and support, and my parents, Rita and Gerald Levy, who are still wondering why anyone would devote years to an endeavor that promises little pecuniary reward. Finally, I want to thank my son, Tal, who is himself becoming more interested in environmental policy and technology. Tal, this book is dedicated to you—you carry the torch of hope for the next generation.

Contributors

Peter Andrée is a Ph.D. Student in the Faculty of Environmental Studies at York University in Toronto, Canada. His dissertation is tentatively entitled "The Global Politics of Agricultural Biotechnology: Canada and the Cartagena Protocol on Biosafety." His research interests include agricultural sustainability, community economic development, and environmental politics. Peter currently teaches in the Environmental and Resource Studies Department at Trent University in Peterborough, Canada.

Jennifer Clapp is Associate Professor in the International Development and Environmental and Resource Studies Programs at Trent University, Canada. She is author of *Adjustment and Agriculture in Africa: Farmers, the State, and the World Bank in Guinea* (London: Macmillan Press, 1997) and *Toxic Exports: The Transfer of Hazardous Wastes from Rich to Poor Countries* (Ithaca: Cornell University Press, 2001). Jennifer is an associate editor of the journal, *Global Environmental Politics*.

David Coen is Senior Lecturer in Public Policy and Director of the MSc in Public Policy in the School of Public Policy at University College London. Prior to joining UCL he was Research Fellow (1997–2001) at the London Business School, where he codirected an Anglo-German study with the Max Planck Group in Bonn on Business Perspectives to Utility Regulation. He is currently editor of the UCL Public Policy Working Paper Series, serves on the editorial board of Business Strategy Review, and is the Chair of the Business and Government Research Network of the International Political Studies Association. David has published extensively and consults on European public policy, business-government relations, and regulatory reform.

Peter Dauvergne is a Canada Research Chair in Global Environmental Politics and Associate Professor of Political Science, University of British Columbia. His research focuses on the international political economy of sustainable development, environmental ethics, and the politics of environmental management in the Asia-Pacific. His most recent book is *Loggers and Degradation in the Asia-Pacific* (Cambridge University Press, 2001). Peter is a member of the Publications Board of the UBC Press and the founding and current editor of the MIT journal *Global Environmental Politics*.

Robert Falkner is lecturer in international relations at the London School of Economics and associate fellow of the Sustainable Development Program at the Royal Institute of International Affairs. Before joining the LSE in 2002, he taught international relations at the universities of Munich, Oxford, Kent, and Essex. His research interests are in IPE and global environmental politics. Robert is co-editor of *The Cartagena Protocol on Biosafety: Reconciling Trade in Biotechnology with Environment and Development?* (2002).

Matthias Finger, Ph.D., Political Science, Ph.D., Adult Education (both University of Geneva) is currently Chair and Professor of Management of Network Industries as well as Dean of the School of Continuing Education at the Swiss Federal Institute of Technology (EPFL). His work focuses on the liberalization of the main network industy sectors—postal services, telecommunications, energy, public transport, water, and air transport—and on issues of regulation and public service.

Lucy Ford is Senior Lecturer in International Relations at Oxford Brookes University. Her research interests are in the areas of International Relations Theory and International Political Economy with particular regard to, on the one hand, governance, civil society and social movements, and on the other, to the political economy of sustainable development. Lucy has published various articles including a coauthored piece with Marc Williams (1999) "The WTO, Social Movements and Global Environmental Management" in *Environmental Politics* 8:1.

Sverker C. Jagers is lecturer and researcher at the Department of Political Science, Gothenburg University. He is also directing its interdisciplinary masters program in social environmental science. Sverker's areas of specialization are green political theory, international environmental politics, and interdisciplinary research methodology. Currently, he is leading a research project on "Public opinions of climate change-related measures and policies."

David Levy is Professor of Management at the University of Massachusetts, Boston. His research examines the intersection of business strategy, technology, and politics in the international arena. In the last few years, he has studied the engagement of U.S. and European multinationals with the emerging climate change regime. He has undertaken research projects in cooperation with the OECD, the UN Center for Transnational Corporations, and the U.S. EPA. David is currently researching the potential of the renewable energy business cluster in Massachusetts.

Peter Newell is a Fellow at the Institute of Development Studies University of Sussex. He is author of *Climate for Change: Non-State Actors and the Global Politics of the Greenhouse* (CUP 2000), coauthor of *The Effectiveness of EU Environmental Policy* (MacMillan 2000), coeditor of *Development and the Challenge of Globalization* (ITDG 2002), and author of many journal articles and book chapters on the political economy of environmental politics.

Matthew Paterson has research interests in the politics of global warming, ecological perspectives in International Relations theory, and the political economy of global environmental change. His main publications are *Global Warming and Global Politics* (1996) and *Understanding Global Environmental Politics: Domination, Accumulation, Resistance* (2000). He is currently working on a book on the political economy and cultural politics of the car in relation to global environmental politics. Matthew is Senior Lecturer in International Relations at Keele University, United Kingdom.

Johannes Stripple holds a licentiate of Philosophy in Environmental Science from Kalmar University, Sweden, and is now a Ph.D. student in Political Science at Lund University, Sweden. Johannes' research concerns the intersection of global environmental change and international relations, in particular questions regarding security, territory, and authority. Johannes' main publications are "Climate Change as a Security Issue" (in E. Page and M. Redclift, *Human Security and the Environment*, Edward Elgar 2002), and *Climate Governance beyond the State* (Global Governance 9, 2003, with Sverker Jagers).

The Business of Global Environmental Governance

1

Introduction: The Business of Global Environmental Governance

David L. Levy and Peter J. Newell

Business plays a key role in international environmental politics. Private firms are engaged, directly or indirectly, in the lion's share of the resource depletion, energy use, and hazardous emissions that generate environmental concerns. Corporate activity dominates every stage of the value chain, from research into genetically engineered food and seeds, to the disposal of household and industrial waste. At the same time, firms can also serve as powerful engines of change, who could potentially redirect their substantial financial, technological, and organizational resources toward addressing environmental concerns. The environmental impact of firms' activities makes them central players in societal responses to environmental issues. According to the Business Council for Sustainable Development's (BCSD) own figures, "Industry accounts for more than one-third of energy consumed worldwide and uses more energy than any other end-user in industrialized and newly industrializing economies" (Schmidheiny 1992, 43).

In many ways, large firms are the "street-level bureaucrats" of environmental policy, Lipsky's (1980) term to describe the role of front-line employees in shaping policy through its implementation on the ground. The active cooperation of large multinational companies is therefore key to the implementation of environmental regulations and the amelioration of environmental problems. Industry appears to be increasingly aware of its role. The International Chamber of Commerce, an influential umbrella industry association, has forcefully asserted industry's significance in the case of climate change, though these words would apply equally well to many other environmental issues:

Industry's involvement is a critical factor in the policy deliberations relating to climate change. It is industry that will meet the growing demands of consumers for goods and services. It is industry that develops and disseminates most of the world's technology. It is industry and the private financial community that marshal most of the financial resources that fund the world's economic growth. It is industry that develops, finances and manages most of the investments that enhance and protect the environment. It is industry, therefore, that will be called upon to implement and finance a substantial part of governments' climate change policies. (International Chamber of Commerce 1995)

This statement by the ICC acknowledges that the natural environment is closely intertwined with economic, financial, and industrial activity, and positions industry as the resourceful, benevolent provider of technological expertise and environmental goods and services—a solution rather than a threat to the environment. Where the scope and impact of private economic activity has traditionally been seen as the subject of regulation, it is now being invoked to justify industry's active agency in the development and enforcement of policy. Above all, industry is asserting its political role in the environmental policy realm.

This book proposes a political economy approach to understanding the role of business in international environmental governance. Despite increasing acknowledgement in the rapidly expanding literature on global environmental politics of the important role that industry has played in the negotiation and implementation of individual regime arrangements, we continue to lack both an understanding of the diverse ways in which firms contribute to the overall architecture of global environmental governance and a sophisticated comprehension of the reciprocal relationship between corporate strategy and international environmental regulation. The aim of this book, therefore, is to develop thinking about the ways in which business activity is both a response to, as well as constitutive of, environmental governance at the global level.

We use the term "environmental governance" to signify the broad range of political, economic, and social structures and processes that shape and constrain actors' behavior towards the environment. Environmental governance thus refers to the multiple channels through which human impacts on the natural environment are ordered and regulated. It implies rule creation, institution-building, and monitoring and enforcement. But it also implies a soft infrastructure of norms, expectations, and social understandings of acceptable behavior towards the environment, in processes that engage the participation of a broad range of stakeholders. We avoid the term "regime" here because, as explored later in this chapter, the concept of environmental regimes is already well developed in the International Relations literature and therefore carries with it a particular connotation of a formal international institution. From our perspective, the main limitation of the concept is that, while it increasingly recognizes the significance of private actors and informal, normative structures, it is still primarily concerned with official inter-state and supranational arrangements. Thus, regime theorists can talk about the success of establishing an international regime to control ozone-depleting substances, or the absence of an international regime to address tropical logging. In contrast, we use the term environmental governance to suggest a broader system of order and structure, one that is always present. In some cases, it might be a purely market-based form of governance, in which environmental impacts flow from private firms as they choose which products to develop, manufacture, and sell. In other cases, industry associations might promulgate their own sets of standards. Even when there is no direct governmental regulation of environmental impacts, patterns of research and production are structured and mediated by systems of property rights and market institutions, by norms and laws that regulate trade and investment, and by the strategic interaction of firms in competition for markets and resources within specific industry structures. One implication is that while international regimes always modify structures and processes of governance in particular ways, they do not necessarily "solve" environmental problems.

A focus on the political economy of environmental governance suggests that attention needs to be paid to the interactions between politics and economics. This demands, first and foremost, that we take seriously the role of the firm as a political actor. Business has been involved throughout the history of environmental policy, especially at the national and subnational level, though it has traditionally been perceived as a more passive actor. It is becoming more apparent, however, that firms are now key political players, engaging with and shaping global processes in direct and indirect ways. Individually, as well as through sectoral and issue-specific organizations, firms are working at national, regional, and international levels to influence policy on prominent environmental issues. As they gain experience, firms are evolving sophisticated strategies to ensure that their interests are represented, and to varying extents, accommodated. Of course, even large multinational corporations are not omnipotent, and are unable to dictate environmental policies. Nevertheless, the evidence in the cases in this book and elsewhere suggests that firms and industry associations do have substantial influence. In a few, perhaps overemphasized examples, such as DuPont in the case of ozone-depleting chemicals, and Exxon-Mobil in the case of greenhouse gases, a single company has been instrumental in shaping U.S. policy, with direct consequences for the direction of the Montreal and Kyoto protocols. More generally, governmental negotiating positions on international treaties tend to track the stances of major domestic industries active on each issue.

An examination of the firm as a political actor needs to extend beyond traditional activities such as lobbying and donations to election campaigns. In the negotiation of many international regimes, business has a formal voice in advisory panels and in the process of authoring and reviewing scientific reports. In the climate change case, for example, the contribution of business to the scientific evaluation process was significantly expanded in the Third Assessment Report of the Intergovernmental Panel on Climate Change. The broader view of environmental governance adopted here suggests that more market-oriented corporate activities can also be viewed as political. In the ozone case, the technological strategies of leading chemical companies drove their stances toward international regulation and helped to shape the content, timeline, and implementation of the resultant protocol. Deregulation of electric power markets, privatization of water supply, and even industry consolidation through mergers and acquisitions, shift the market structures and competitive dynamics that shape business operations, often with significant environmental consequences. Fundamentally, the legitimacy of business activities is a deeply political issue, and activities directed toward sustaining this legitimacy in the face of regulatory pressure and public distrust should be understood in this context. The language and practices of environmental management, for example, from the adoption of environmental standards such as ISO 14000 to the development of partnerships with environmental nongovernmental organizations (NGOs), are rapidly permeating large corporations. In a political economy perspective, these activities are more than simple technical and economic techniques and cannot be dismissed as cynical public relations "greenwash"; rather, they represent an integrated political and market response to environmental challenges, in a way that serves to accommodate social and regulatory pressures.

The increasing importance of international environmental agreements for a wide range of industry sectors, together with the significant impact of business activity on both the environment and the systems of governance set up to protect it, suggests that more attention needs to be paid to corporate strategy. Companies facing similar environmental pressures often adopt radically different strategies, ranging from strong opposition and challenges to the scientific basis for action, to constructive engagement and investment in alternative technologies. These differences sometimes defy simple explanation in terms of objective economic interests. If policymakers are to harness and steer corporate resources in environmentally constructive ways, it is of critical importance to examine and understand the determinants of these diverse business strategies. Similarly, business managers need to develop more sophisticated understandings of environmental problems, the potential of various technologies, and trends in the regulatory and market environment, if they are to develop strategies that integrate the pursuit of market opportunities with efforts to reduce the environmental impacts of their activities. One recurring theme in the book is that strategies are not developed on the basis of a set of fixed, objective interests; rather, strategies rest precariously on perceptions of interests that are constructed in institutional environments, and are thus influenced by national and industry contexts. Perceived interests can therefore shift over time, and within the negotiation process itself.

A political economy approach to international environmental governance also promises to enrich our understanding of global environmental politics. Regime theory, with its emphasis on interstate relations, has not paid sufficient attention to the rich and complex process of political bargaining and negotiation among a range of actors, most importantly firms, industry associations, NGOs, state agencies, and international organizations (which can also serve as a forum for these negotiations). This process includes efforts to deploy scientific and economic assessments to frame debates in particular ways, to forge broad alliances, and to project a specific conception of the general interest.

Regime theory has also tended to portray regimes as rational, technical solutions to environmental problems that successfully overcome problems of collective action among states. A political economy perspective, however, emphasizes that the structures and processes of regimes effect asymmetrical distributional outcomes, not just for states, but also for various industrial sectors and other social groups. Indeed, an observer of many complex, protracted negotiations could easily be forgiven for concluding that these distributional impacts are far more important to participants than any amelioration of the environment. The policy process apportions responsibilities, environmental risks and benefits, and distributes the burden of action. Developments in the mode of environmental governance will therefore create differential impacts on trade and investment, prices and profits, and employment and wages. New norms and forms of discourse will become accepted and established; indeed, the balance of power between business, state agencies, environmentalists, and other social groups will itself shift.

Approaching environmental governance from a political economy perspective provides some traction for analyzing the impact of broader patterns and trends in the global polity. Traditional regime approaches tend to underplay the significance of global forces that lie outside the particular institutional arrangement under scrutiny. Deals brokered at the international level reflect compromises and trade-offs negotiated over time with a range of key actors. The degree to which different actors are influential in such deal-brokering, and the repertoire of arrangements considered legitimate, requires an understanding of power as it is exercised, not just "within" a regime, but in multiple political, economic and social sites within the global political economy. Perhaps the most important trend, in this context, is the strengthening of institutions and the diffusion of discourses associated with neoliberalism, including elements such as privatization, market pricing, extension of property rights, and the removal of restrictions on trade and investment. These are common elements in most of the regimes examined in this volume. The privatization of governance is most obvious, perhaps, in the development of private regimes, such as the environmental management standards ISO 14000 (Clapp, this volume). Our broader perspective on governance, however, suggests that other phenomena can also be interpreted as a form of privatization of governance. Quasi-private policy bodies such as the Trans-Atlantic Business Dialogue are becoming increasingly influential in trade and investment policy, as well as standard setting (Coen, this volume). Entire industries, such as water supply, are being privatized (Finger, this volume), while private firms are increasingly responsible for developing new technologies, environmental monitoring and assessment, and provision of financing. Trading mechanisms are being established for greenhouse gas emission rights and for bonds that bear the risk of property damage from severe weather events.

While the broader political and economic context is clearly important, our approach does not accord it a structural, deterministic power over particular environmental modes of governance. The cases in this book suggest that governance arrangements evolve over an extended period of complex, multiparty negotiations. Outcomes are influenced by the specific skills and strategies of the various actors as they try to build coalitions, frame issues in particular ways, and influence decisions makers. By their nature, strategies involve risks and the possibility of failure. Each issue has its own competitive dynamics, regulatory and institutional context. Chance coincidences, such as the scientific testimony about climate change before the U.S. House Energy Committee on a record-breaking hot week in Washington, D.C., in June 1988, can create strategic windows of opportunity. Outcomes thus exhibit some degree of indeterminacy. Moreover, the outcomes from any particular set of negotiations influence the norms, practices, and institutions that will affect other issues. Corporate engagement with the ozone case, for example, gave rise to processes for inclusion of business in technical and scientific issues that set the stage for the climate change regime. The relationship between issue-level governance and the global political economy is thus dialectical; individual regimes are shaped by, yet constitutive of, wider political and economic structures.

This Book's Insights and Perspectives

The focus of this book is a detailed theoretical and empirical analysis of the ways in which firms interact with governments, NGOs, and other actors as they both shape and respond to the agenda of international environmental politics. The contributors explore new forms of environmental governance, as well as new business and NGO strategies, that arise as a result. The book not only provides comparative insights into the responses of business to major international environmental issues, including ozone depletion, climate change, and genetically modified organisms (GMOs), but also explores how these responses differ across sectors, regions, and issues, and how they evolve over time.

The book also aims to contribute to the development of theoretical frameworks for understanding the role of business in global environmental politics. Scholars from a range of disciplines have been brought together to assess the relevance of existing theoretical tools and the possibilities of alternative approaches. Our contributors come from different disciplines, including organization theory and strategy, international relations, and political science, yet they share an interest in the role of private actors in global environmental politics. It is our intention to draw from this range of disciplines in order to build a conceptual bridge between the microlevel analysis of strategic interactions among firms, governments, NGOs, and international organizations, and more macrolevel analysis of the emerging system for global environmental governance. We also hope to build connections and conversations between two very different academic worlds that rarely encounter one another, those of management and business on the one hand, and international relations and environmental politics on the other.

The theoretical and empirical contributions in this volume represent an advance on the current literature on global environmental politics, which tends to neglect the role of private actors and lacks a coherent account of the ways in which firms are implicated in global governance of the environment. The editors, David Levy and Peter Newell, propose a neo-Gramscian framework that offers a flexible approach to understanding the contested and contingent nature of business power, the complex processes of alliance building and accommodation, the key role of civil society in establishing legitimacy, and the integration of economic, discursive, and organizational strategies. It is an approach that steers between, on one extreme, overly rigid structural accounts that adduce overarching power to multinational capital or dominant states, and, on the other side, pluralist approaches that presume a rough equivalence among actors and neglect the systematic asymmetries that flow from wider political economic structures. The various contributors to this volume do not, however, share a single theoretical perspective. Some see business enjoying a dominant position, while others attach less relevance to the structural dimensions of business power. Some focus more on the interplay of economic and material interests, while others emphasize discursive and cultural interpretations. Nevertheless, the authors concur in adopting a political economy approach, in which interplay of economic structures, corporate strategies, and political processes drive the evolution of international environmental governance. All the contributors view firms as key actors in this process, in their responses to and influence over environmental regulatory mechanisms, and in their responsibility for environmental impacts of economic activity.

Structure of the Book

Part I: Conceptual Frameworks We have noted in this introduction a general neglect of the role of business in global governance. Chapter 2 explores in more detail the current "state of the art" of literatures and debates about the firm as a global actor. It provides an overview of the various strands of regime theory, and notes some of the limitations of conventional approaches to international environmental governance. The chapter examines some recent theoretical developments, reviews debates about various forms of privatization of authority, and explores the changing nature of the relationship between business, the state, and civil society in a context of globalization. The review helps to lay the ground for a more coherent political economy approach that builds on regime theory to give much more prominence to economic structures and corporate strategies.

In chapter 3 we construct a neo-Gramscian conceptual framework, which proposes that international environmental governance is a profoundly political process that engages business, NGOs, state agencies, and intergovernmental actors in contestation over structures and processes of governance. This framework constitutes an effort to explain and illuminate the developments in the various environmental arenas discussed in subsequent chapters. The framework attempts to integrate perspectives from International Relations and International Political Economy with approaches from corporate strategy and environmental management. Our approach emphasizes the contested and contingent nature of business power, and that this contestation takes place across multiple levels (regional, national, international) and sites of power (material, discursive, organizational). This approach suggests a broad notion of environmental governance and its close relationship with industrial structures and business dynamics. The framework highlights the political nature of efforts to protect market position and build social legitimacy, through the use of a range of strategies including technological innovation, the construction of coalitions, and engagement in public debates over the science and economics of environmental issues.

Part II: Business Strategies and International Environmental Governance
While there is a growing acknowledgement of the importance of nonstate actors in global environmental politics, few analyses place firms and corporate strategy centrally in their analysis of international environmental governance. The chapters in this section, taking three high-profile issues as their subject matter, do just that.

In the first chapter in this section, David Levy examines different corporate response strategies in Europe and the United States to climate change, and argues that economic and market structures only partially explain the greater tolerance in Europe for carbon emission limitations. Given the high level of uncertainty concerning science, technology, and policy, he contends that companies have a degree of discretion in their responses, and are therefore influenced by their institutional environments. In the early years of an issue, the influence of the home country environment tends to predominate, but as issues mature, he argues that firms are increasingly influenced by the global industry and by the institutional apparatus of environmental issues, leading to convergence in corporate strategic responses.

Drawing from the neo-Gramscian theoretical framework, the paper argues that the evolution of business responses to climate change can be understood in terms of industry's efforts to sustain its hegemonic position in the face of challenges. The sea change in the stance of U.S. business toward climate change that occurred in the period 1996 to 2000 was, he argues, a result of an accumulation of changes in technological capabilities and economic opportunities, organizational forms, and discursive structures. Fundamentally, the fossil fuel industry failed to secure legitimacy for its position in the key realm of civil society; we therefore observed a strategic shift toward an accommodation, hence protecting both corporate legitimacy and key economic interests of major sectors. The resulting regime is weak and modest, and is increasingly driven by fragmented decentralized efforts rather than a formal international treaty. While this compromise might construct an alignment of interests sufficient to bind a loose coalition of actors together, it may well be inadequate in terms of environmental protection.

The second chapter in this section, by Robert Falkner, posits that the technological resources of private firms constituted a significant source of their power in international ozone politics. Though the ozone story has been told many times from different perspectives, Falkner's chapter advances our understanding of corporate involvement in international ozone politics in several ways. While most studies emphasize the role of the major CFC producers, Falkner highlights the role of the CFC user industries in the unfolding negotiations. He shows how differences between and within producer and user groups in the major CFC industry sectors have influenced the process from the birth of the ozone regime through to its contemporary structure. The story is explained in terms of "business conflict" between competing sectors, highlighting the role of technology as a source of power for the firms, and the centrality of business interests in guiding state bargaining positions. He concludes that while business was a key player in shaping the form and implementation of the Montreal Protocol, business was not "in control" and the agreement still represented an unwanted compromise.

The third chapter, by Peter Andrée, invokes the neo-Gramscian framework to examine how the major biotechnology companies have been involved in a "war of position" with environmental groups and state regulatory agencies. Despite strong government support for biotechnology and vast levels of investment by biotechnology companies in the development of GM foods, activists, particularly in Europe, have successfully persuaded consumers and some regulators to reject biotech products. Andrée describes the strategies that the industry has used to construct support for the technology and deflect challenges to the hegemonic bloc they have been seeking to maintain. He looks at the material, discursive and organizational underpinnings of the biotech bloc, providing a sense of how corporate strategies have evolved over time and continue to adapt to the shifting terrain of contemporary biotechnology policy. These shifts are related to broader patterns of change in the global political economy that are central, in Andrée's view, to understanding the global politics of biotechnology.

Part III: Business Influence: Regional Dimensions   There is a danger inherent in any attempt to generalize assumptions about the ways in which firms organize and represent their interests. Chapter 4 of this book by Levy, and other work by Levy and Newell (2000), points to important differences in the strategies and lobbying tactics adopted by firms; these differences may be a product of the political structures or social and cultural values of the region in which firms are based. Of course, many of these firms are multinational corporations, operating across multiple jurisdictions and cultures, while also needing to respond to emerging multilateral sources of authority, such as the institutions of international environmental regimes. Locating corporate engagement with environmental governance systems in a regional context and probing the interconnection of national and international structures provides a richer picture of business responses to environmental regimes.

Peter Dauvergne's chapter examines the influence of environmental pressures on the actions of tropical logging companies in the Asia-Pacific region. He illustrates that over the last decade, environmental ideas, agencies, and activists have helped push the rhetoric and policies of governments and firms to become more sensitive to environmental concerns. While some important shifts have developed, however, few concrete changes have occurred so far in corporate environmental practices on the ground. Dauvergne highlights the disjuncture between the formal and informal aspects of the logging regime, characterized by the gulf between negotiations, laws and policies on the one hand, and customs, norms, and patronage networks on the other. Added to this, he describes how region-specific networks of corporations, financiers, managers, state officials, and traders reduce corporate accountability and transparency, making it difficult for environmental activists and states agencies to influence corporate practices. Dauvergne's account underscores the importance of looking at the local social, political, and economic systems of which firms are a part in order to understand how well-intentioned global environmental initiatives can be subverted and undermined at the regional and national level. It also highlights the significance of the broader concept of governance developed in this volume, as environmental impacts are driven by local economic imperatives and political processes.

Regional influences have a bearing not only on corporate market strategies but also on their political, or non-market strategies. Europe has seen a boom in economic and public interest lobbying focused on institutions of the European Union, reflecting shifts in decision-making power from the member states to the EU. Picking up on a trend towards trans-Atlantic convergence noted in Levy's chapter, David Coen's chapter explores current patterns of business-government relations in the EU and the United States, and observes that, from very different starting points, there is a measure of convergence in trans-Atlantic corporate political strategies, even if important differences remain. In particular, he notes the growth of issue-specific industry associations that attempt to represent the interests of firms from multiple sectors and countries, and which operate across multiple levels of decision-making. These trends are explored with examples from the regulation of automobile emissions.

Part IV: The Privatization of Governance: Business and Civil Society
Rather than accede to government-set standards, many business sectors have taken the initiative in setting up their own regimes for certification and standardization. In the environmental arena, codes of conduct such as the Valdez principles have proliferated, as have stewardship regimes that accredit responsible environmental practice and establish standards

and labeling schemes. These business initiatives, often accomplished in partnership with NGOs, represent an increasingly important component of the global environmental architecture. Some of these efforts relate to the environmental performance of consumer products, while some attest to the conditions of production along the supply chain. Those firms that have invested in cleaner products and processes increasingly demand recognition for their efforts, and private certification regimes enable firms to develop brand recognition and deter competition, while deflecting demands for governmental regulation. Contributions in this section highlight the various mechanisms by which private actors establish private regimes of governance, and point to some problematic implications regarding participation in governance processes and the distribution of costs and benefits.

Jennifer Clapp's chapter in this volume explores the environmental management systems endorsed by the ISO. The ISO 14000 standards are being adopted by standards setting bodies in some states as national EMS standards, and are now recognized by the World Trade Organization (WTO) as legitimate public standards and guidelines. Clapp shows how the growth of private standard-setting bodies has led to mixed regimes of a hybrid nature, whereby both states and private authorities are heavily involved in the creation and maintenance of international principles, norms, rules, and decision-making procedures. Clapp explores the consequences of such private standard-setting processes, where membership and procedures are often far from open and participatory. The ISO 14000 standards are of particular concern for developing countries that do not have as much representation in the body as do industrialized countries, as well as for smaller firms, for whom the standards can constitute a market barrier. Clapp thus highlights both the political and ecological consequences of this shift in environmental governance towards the privatization of authority.

While we are accustomed to thinking of private governance in terms of self-regulation by firms, Sverker C. Jagers, Matthew Paterson, and Johannes Stripple, show how the insurance industry is creating its own set of private governance mechanisms through what they refer to as the securitization of risks from climate change. They demonstrate that, while many insurers have become concerned about the possible impacts of global warming on their industry, the dominant response by insurers has been to develop new financial instruments that extend their capacity to finance losses from large-scale weather-related disasters. This is in spite of the attempts by individuals within Greenpeace to divide business opposition to action on climate change by forming an alliance with the insurance industry. These moves are understood by the authors in Gramscian terms as a counterhegemonic strategy to fracture the power of the fossil fuel industry, by preventing it from representing itself as *the* broad voice of business. The authors explain the relative failure of this initiative in terms of financial interests and the political culture of the insurance industry. They demonstrate how the risk management strategies of insurance companies constitute a new form of private governance that avoids the need to form alliances with environmentalists. These strategies, moreover, have far-reaching consequences for those subject to the risks of climate change, particularly the poor and vulnerable in less developed countries, who are often left uninsured. The asymmetrical access to and outcomes of these governance systems highlight the importance of paying attention to the political economy of private forms of governance.

The shift towards the privatization of authority in the various guises described in the previous chapters is ultimately validated by a belief in the superiority of market-driven policy. In many national settings, this ideology is promoted by institutions often not thought to be part of the system of global environmental governance, such as the World Bank. Yet the development projects and programs that such institutions finance and support have far-reaching social and environmental implications.

In his chapter on the privatization of water supply, Matthias Finger highlights the importance of relations between firms and international development and finance organizations in understanding the privatization of water and the implications for questions of access and water quality. Finger explains the background to the increasing popularity of what he labels the "new water paradigm," in which private firms and market pricing form the backbone for water supply systems. This water paradigm is an example of a loose environmental governance system governed by neoliberal principles and dominated by a handful of water multinationals and international financial organizations. Having profited from the World Bank's privatization policies and successfully acquired water companies in developing countries, multinationals in the water industry are now turning towards Europe and the United States. As noted in other chapters, the privatization of governance often marginalizes the influence of NGOs and other social groups, leading to detrimental distributional consequences. The paper notes a tension between patterns of liberalization and reregulation; although states are providing fewer services with diminished resources, water multinationals are lobbying for tougher environmental standards that only they are in a position to meet. Through these means, it is claimed, firms are bringing about the "instrumentalization" of the state in order to obtain markets, financing, and secure property rights. For Finger, the benefits of a Gramscian approach relate to the significance of the ideology of the "new water paradigm," and insights gained into the process of coalition-building between firms, NGOs and international institutions through emphasis on public-private partnerships.

The ways in which business engages with institutions of global environmental governance and the extent to which they adapt corporate strategies to environmental initiatives is increasingly affected by the activities of civil society groups. Civil society is an important battleground, therefore, for broader social and political conflicts. In the final chapter of the book, Lucy Ford draws on the experience of the toxic waste trade to explore how NGOs and social movements have sought, in different ways, to interact with and engage the regime set up to govern the global trade in toxic wastes. While some groups have sought to "play by the rules," working with policymakers and using prevailing discourses to lobby for change, others have questioned predominant forms of scientific and expert knowledge used to understand environmental change and to raise more fundamental questions about production, consumption, and lifestyle that are neglected in mainstream policy discourses. The chapter provides an interesting insight into the strategic dilemmas facing NGOs: the dangers of cooptation by dominant groups versus the risk of marginalization that comes with nonengagement. Ford develops a neo-Gramscian position to challenge the conventional view that civil society always represents a democratic force in global politics, and cautions that civil society is also a key site of political struggle for legitimacy, allies, and influence.

We conclude the book by drawing out common themes, insights, and experiences from the sectors, issues and regions explored in the chapters. We suggest what these insights might add to our understanding of the role of business in international environmental governance, and reflect on the theoretical framework in light of the cases presented. We suggest that one does not need to refer to hegemony or historical blocs to reach many of the key insights of this volume, such as the pervasive influence of business, the legitimating role of civil society, or the significance of business strategies. However, the neo-Gramscian framework integrates these insights in a more theoretically grounded and intellectually satis-fying manner, providing a more systemic understanding of dynamic processes of political contestation over environmental governance, and their linkages to macropolitical and economic structures. More gener-ally, a political economy approach enables us to conceptualize the devel-opment of systems of environmental governance not just as a rational problem-solving activity, but as a political effort to coalesce an alliance of groups around a specific set of arrangements. This conception allows us to ask questions about the distributional and environmental impacts of different governance mechanisms. Moreover, an understanding of the strategic dimension of power implicit in the political process opens ana-lytical space for environmentalists and other actors to consider the pos-sibilities for advancing or contesting particular forms of governance, to enhance protection of the environment.

References

International Chamber of Commerce. (1995). Statement by the International Chamber of Commerce before COP1, March 29, Berlin.

Levy, D. L., & Newell, P. (2000). Oceans apart? Business responses to the envi-ronment in Europe and North America. *Environment, 42*(9), 8–20.

Lipsky, M. (1980). *Street-level bureaucracy: Dilemmas of the individual in public services.* New York: Russell Sage Foundation.

Schmidheiny, S. (1992). *Changing course.* Cambridge, MA: MIT Press.

I

Conceptual Frameworks

2

Business and International Environmental Governance: The State of the Art

Peter J. Newell

The purpose of this chapter is to provide an overview of the conceptual and theoretical tools that have been used to date to account for the role of business in international governance. We noted in the introduction a general neglect in many mainstream approaches of the role of businesses in global governance. The first part of this chapter elucidates this claim in relation to conventional thinking within International Relations in general, and in relation to the environment in particular. The second part looks at some interesting theoretical developments in relation to the role of the firm in global governance. It reviews debates about private regimes and the privatization of authority, as well as work that seeks to explain differences in the ways firms respond to environmental challenges and organize themselves to influence policy in different parts of the world. The third part of the chapter looks at how globalization has changed the nature of environmental governance and the role of business in particular. It looks at work on business regulation and the changing nature of the relationship between business and civil society in a context of globalization. This review helps to lay the ground work for an alternative political economy account, the contours of which are developed in the next chapter.

The State of the Art in International Relations: Regimes

International Relations (IR) as a discipline has, on the whole, neglected the role of business in international affairs. As Cox notes: 'The dominance of the statist paradigm has meant that the power and influence of business has often been minimized or ignored in the study of international politics' (Cox 1996, 1). Insights from the community power debates in Political Science about the role of firms in the policy process have not, on the whole, been applied at the international level (Crenson 1971).[1] Instead, an increasingly prominent substream of IR, International Political Economy (IPE), has developed, seeking to emphasize the importance of firms in the global economy (Eden 1991, 1993), often focusing on their relations with states, epitomized by Stopford and Strange's (1991) book *Rival States, Rival Firms*. Strange's (1994b) model of "triangular diplomacy" that describes the triangle of relations between states, between states and firms and among firms provides a useful way of understanding many dimensions of global bargaining, even if has less to say about the role of civil society actors central to global environmental politics.

Given that IPE has placed state-market relations at the heart of its enquiry (Strange 1994a; Stubbs and Underhill 1994), we would expect there to be greater attention to the role of the firm in global governance. Some authors have charted the significance of the rise of nonstate actors in the global system in general terms (Higgott, Underhill, and Bieler 1999), or sought to explore the role of a particular group of business actors in making global public policy (Sell 1999, 2003) or in exporting particular models of environmental regulation (Garcia-Johnson 2000). But despite studies on private regimes that challenge our thinking about the role of firms in global governance formation, the impact of such thinking on mainstream IPE appears to be relatively weak. IPE scholars have also tended to neglect environmental issues. Indeed, it is more the case that writers on the environment have sought to make use of concepts and debates in IPE to account for the global politics of the environment (Newell 2000a; Paterson 2001; Saurin 1996; Stevis and Assetto 2001; Williams 1996). Only rarely have scholars of IPE sought to understand the significance of environmental issues for mainstream theory (Helleiner 1996), even if there has been an attempt to use Green political theory to challenge conventional thinking within International Relations (Laferrière 1996; Laferrière and Stoett 1999).

Insights from IPE on these issues have yet to capture the imagination of much mainstream thinking in IR, however, which continues with a relatively state-centric reading of questions of world order and international cooperation, even if the work of Rosenau and others on "governance without government" has acknowledged that the state is but one actor among many in the global system (Rosenau and Czempiel 1992). While Realism is ambivalent about the role of business, in regime theory business becomes significant only insofar as these groups influence regime formation, maintenance, or disintegration (Nowell 1996).

Within the study of international environmental politics, regime approaches continue to enjoy a privileged status. Regime theory concerns itself with "norms, rules, principles and decision-making procedures around which actors' expectations converge in a given area of international relations" (Krasner 1983, 2) and grew out of a concern that, with the decline in the hegemonic power of the United States, the prospects of international cooperation would be detrimentally affected (Keohane 1984). Hopes turned instead to the ways in which international institutions would be able to regulate state behavior in the absence of a hegemonic power. Applied to global environmental change and the problematic of managing the global commons, regime theory appears to provide a useful analytical grounding for the conceptualization of such problems (Vogler 1995). Regime theory responds to a number of overlapping concerns that traditionally engage scholars of IR, centering on the need to restrain egoistic state behavior and nurture collective agreement in an anarchic international society. The appeal of regime theory for scholars of the environment endures despite critiques from a range of quarters regarding the static nature of much regime analysis, the state-centricity of the approach (Strange 1983), and its neglect of many of the broader political and economic forces that condition the context in which regimes emerge and evolve and the extent to which they are effective (Newell 2000a). Each of these factors is problematic from the point of view of understanding the role of business actors in global environmental governance.

Despite passing acknowledgement, there is little in the regime literature on the role of business in international environmental politics. Accounts from individuals personally involved in environmental negotiations allude to a powerful role for industry groups in the ozone (Benedick 1991; Oye and Maxwell 1994) or climate change regimes (Mintzer and Leonard 1994), for example, but the acknowledgement of business influence has not, to date, extended to revisiting conventional theoretical assumptions. There have been a number of calls for more study of nonstate actors (Caldwell 1988; Haas, Keohane, and Levy 1993; Haufler 1995), which have not on the whole, been followed up in the regime literature. This is in spite of Young and Von Moltke's (1994) acknowledgment that

> it is critical to deepen our concerns for the pervasive role of non-state actors as players in the processes of regime creation, as the ultimate subjects of many regulations, and as pressure groups in the implementation and operation of regimes . . . No issue-area constitutes a better laboratory in which to study these developments than international environmental affairs (pp. 361–362).

Much of the recent work on non-state actors and global civil society does not engage with IR theory (Cohen and Rai 2000; Edwards and Gaventa 2001) with the exception perhaps of Wapner (1996) and O'Brien et al.'s (2000) study, which uses the idea of "complex multilateralism" to explore, among other things, the impact of the environmental movement on global economic institutions such as the WTO (World Trade Organization) and World Bank. The neglect of nonstate actors within mainstream International Relations is partly a function of the way in which states are conceptualized in regime analysis. States tend to be viewed as unitary actors, autonomous from other social and political agents, which pursue their definitions of national interest based on rational calculations of the costs and benefits associated with a particular course of action. The influence of game theory on the assumptions made by regime scholars is apparent in this regard, where it is taken for granted that state interests are given and can be deduced for the purposes of predicting patterns of cooperation and noncooperation (Ward 1996). However, a strict focus upon the "ken of international bureaucracies and diplomatic bargaining" (Strange 1983, 338) precludes assessment of the way in which interests are contested and constructed through interaction with business and other actors at the national and international level. This neglect is particularly problematic in the area of environmental policy where levels of uncertainty are often high and the demand for expertise, as a result, often pronounced. Companies, as "street-level bureaucrats" (Lipsky 1980) and experts in the technologies and production processes that are the subject of regulation, as well as key employers, become central players in helping states to define appropriate and viable responses to environmental threats.

Some strands of regime theory are more open, however, to the idea that processes of social learning allow for the (re)negotiation of preferences and interests. Social learning is a key role ascribed to international institutions by interest-based approaches to regimes, in so far as they help to generate cooperative solutions that encourage states to see beyond their own interests for the common good (Hasenclever et al. 1997). The exchange of information, the creation of fora that allow for iterative bargaining over time, thereby creating a "shadow of the future," and the establishment of monitoring mechanisms to deter free-riding, are all said to alter the incentive structure in favor of cooperation. It remains the case, however, that interest-based approaches assume learning and bargaining to take place between states rather than states and nonstate actors. The negotiation and accommodation that takes place at international meetings, and which often comes before it, involving a range of nonstate actors, is not considered to require a revision of these basic understandings of how institutions are constructed.

Alternative accounts that employ a more constructivist approach are better placed to capture these dynamics, viewing regimes as products of socially negotiated "regimes of truth," which serve to draw fields of meaning around problem definition, feasible solutions, and legitimate participants in the process (Keeley 1990). Hajer's (1997) work on "discourse coalitions" shows how particular discursive frames help to shape perceptions of interest and thus align networks of actors into coalitions. Litfin's (1994) reading of global ozone politics adds an historical element to such an approach, showing how particular framings of a problem become salient, acceptable and practicable at particular moments in the debate. Emphasizing the social construction of regimes and the importance of the relationship between knowledge and power enhances our ability to locate regime formation as a product of a set of historical circumstances, though the material underpinnings of this process are often lacking in such accounts. Such accounts also help us to understand the contested nature of regimes, which is important for understanding the entry points available to counterhegemonic movements.

Knowledge-based theories of regimes, of which Peter Haas's (1990) work on epistemic communities is the most widely known, might be considered best placed among regime theories to appreciate the role of businesses as knowledge-brokers, constructing what is and is not policy-relevant knowledge, and sponsoring frames of interpretation that advance their own interests in the debate. Much of this work to date, however, focuses on scientific expertise as an "objective" form of knowledge capable of reducing the uncertainty that characterizes environmental problems and building consensus in favor of cooperation. What is often lacking here is attention to the contested nature of knowledge, in which firms (and NGOs) are actively engaged in supporting or challenging particular interpretations of evidence that sustain or subvert the case for action. Economic and technological forms of expertise are particularly important in this regard. For example, economic models regarding the costs and benefits of different courses of action and technological modelling on available technologies and production techniques for combating environmental problems are widely used and clearly involve industry actors in their production. Indeed, as Coen discusses in this volume, corporate expertise is an important currency of political access and influence as well as validating mechanism for government policy choices where the buy-in of key actors is a prerequisite to success (Grant et al. 2000). At a more fundamental level, whose knowledge counts and which forms of knowledge are marginalized are key questions that narrow understandings of what constitutes knowledge tend to overlook.

Moreover, instead of viewing scientists merely as facilitators of co-operation, a political economy approach, building on the work of writers such as Boehmer-Christiansen (1994), would view them as another political interest in the debate, seeking alliances with NGOs and media to advance and entrench their position and engaged in a game of political bargaining which affects the content and presentation of their advice. The political and economic significance of the solutions they are in a position to propose helps to explain the intensity of contests over the construction of policy-relevant scientific findings (Jasanoff 1995; Wynne 1994). A political economy approach would want to account for the reasons why the knowledge forms that businesses often use to substantiate their policy preferences are privileged in policy discourse, relating this both to their deployment of salient modern rational-technical discourses to support their claim-making as well as their material power that helps to ensure that their voice is heard over others.

The potential of regime theory to account for the role of business in global governance is further limited by its acceptance of the separation between domestic and international politics. Given that firms often seek to extend their influence by forging connections between domestic and international fora, this is especially problematic (Coen, this volume; Newell 2000a). Businesses, especially where they form international coalitions, cannot be thought of either as exclusively national or international actors but as both, acting simultaneously across these levels. In this sense, they "transcend the level of analysis problem" (Nye and Keohane 1972, 380). Putnam's (1988) notion of a two-level game (domestic and international) offers a more refined account of the global dynamics of cooperation, in that it offers the possibility of including domestic level analysis: an approach to understanding both the "internationalisation of domestic politics" and the "domesticisation of international politics" (Nye and Keohane 1972, 376). It fails to explore the idea, however, that firms can affect the "win-sets" that are thought to guide state behavior.

Transnationalist approaches, such as that developed by Risse-Kappen (1995), which attempt to get beyond the state-centrism of regime analysis and look at how global political outcomes are mediated through domestic institutions and coalitions, may help to advance a theory of regimes that is able to respond to some of these criticisms. Especially where emphasis is placed on the formation of transnational coalitions of interest that take the form of strategic alliances on particular issues, or a more settled form of policy network, analysis is able to transcend state/nonstate and domestic/international analytical divides. Attention is instead focused upon the internationalization of bureaucratic politics, where internal divisions between government departments get reproduced on the global stage and government actors seek support and inputs from nonstate actors to bolster their positions. Such alliance building involves the corporate sector, NGOs and the scientific community, each trading on the political assets and status that they are able to wield in the policy debate. Jakobsen (1997) and Newell (2000a) have sought to develop these approaches in understanding global climate politics. In addition to emphasizing the plurality of actors and the multiple levels across which power is exercised, they adopt a political economy framework to account for why some coalitions and alliances of interest have a greater bearing on global policy than others. Policy responses and the governance structures that give rise to them are understood as products of a particular configuration of historical and material circumstances.

Hence while some strands of regime theory are better placed than others to provide insights into the role of business in global governance, many fundamental assumptions that underpin the theories serve to weaken their ability to take business seriously. Attempts to evolve regime theory in new directions and to build on critiques of mainstream approaches may yet help to provide a pluralist account of global governance in which business actors are merely one among many, perhaps further developing the idea of "complex multilateralism" or re-invigorating debates about "transnationalism." Challenging classic critiques of regime theory, such as that of Strange (1983), Gale (1998) has suggested that, stripped of its neorealist and neoliberal heritage, the concept of a regime could even be deployed within a critical neo-Gramscian perspective of the sort we develop in the next chapter. He suggests that such an account broadens our understanding of how power is exercised within a regime, forcing us "to widen our focus beyond the diplomats who are formally engaged in negotiations to include the struggles taking place among competing social forces over the principles, norms, rules and procedures of the international regime" (Gale 1998, 277). This perspective simultaneously seeks to correct the state-centrism of mainstream approaches to regimes, emphasize the key role of non-state actors, draw attention to the ideological dimension of international regimes, and help to transgress distinctions between national and international politics by looking at the way in which "regimes are maintained and transformed as a result of the changing balance of social forces within, between and above states" (Gave 1998, 277). The next chapter explores in more detail how a political economy account emphasizing a combination of material, institutional, and discursive power, as well as an analysis of the historical and material underpinnings of particular coalitions of interest, can add to our understanding of the function of business in global governance.

Privatization? Private Regimes and Business Regulation

Partly as a response to some of these criticisms of regime theory, but also as a result of the undeniable growth of industry's own attempts to construct regime arrangements to facilitate commercial transactions, an interest in "private regimes" has recently developed (Cutler et al. 1999; Haufler 1995). Private regimes are international institutional mechanisms, aimed at bringing order to an area of business activity, in which state authority is either not present at all, or not the predominant form of political authority. Private environmental governance presents one manifestation of this trend (Falkner 2003). Prominent examples in the environmental area would include the ISO (International Organization for Standardization) standards, or the FSC (Forestry Stewardship Council) and MSC (Marine Stewardship Council) certification regimes. They can be understood as an expression of a perceived need on the part of industry to enjoy the benefits of a regime, in terms of establishing rules, norms and decision-making procedures, without requiring the exercise of state power to validate and enforce them.

In many cases, enforcement takes place through reporting, auditing, and inspection by other private authorities. Sanctions include withdrawal of a product from certification and, more importantly, loss of public reputation, and the corresponding financial implications of damage to a brand name. Not bound by the same customs and expectations as state-based regimes, with their requirements for consultation with civil society, equal rights to representation and transparency of proceedings, private regimes allow for faster-track decision making, where actors involved are driven by the commercial gains to be made from product endorsement, reduced transaction costs, and access to markets. The growing significance of the rules and standards that are set by these bodies for public regimes such as the WTO, for other market actors and for the overall quality of environmental standards, means that they will undoubtedly attract further academic attention (Finger and Tamiotti 1999; Krut and Gleckmann 1998). Not only do they indicate the ability of firms to create their own regimes, which affect the overall structure of global environmental governance, but at a deeper level, they imply the willingness of business to "venue shop" for the most appropriate forum for governance.

It is not just the case that the relationship between public and private regimes is affected by the role of firms in international environmental governance; the balance of power and authority among public regimes may also be altered by the ties that institutions maintain with corporations. For instance, the close interactions between trade officials and firms in cases before the WTO dispute settlement panel and in the drafting of accords such as TRIPs (said to have been heavily driven by the copyright concerns of the U.S. software industry) are well known (Sell 1999). But when conflicts emerge between the provisions in one regime and another, business may play a crucial role in persuading states to back one set of rules over the other. This is clearly the case in the ongoing debate over the extent to which the Biosafety Protocol is subordinate to, or overrides WTO rules (Newell and MacKenzie 2000). Industry coalitions have played an active role in lobbying for provisions that minimize the opportunity for restrictions on trade. The outcome of these battles, where the rules of regimes appear to pull in different directions, will be shaped to a significant degree by industry actors, active in both environmental and trade and investment regimes, working to ensure that neither impinges on their commercial ambitions (Newell 2003a).

There has also been increasing attention to the role of firms within the United Nations system as a whole. This interest reflects and coincides with the heightened involvement of firms in bodies such as the Codex Alimentarius Commission and in standard-setting bodies dealing with health and environmental issues (Lee, Humphreys, and Pugh 1998). The trade implications of these standards, plus the expertise that manufacturers and exporters bring to these discussions, helps to explain the high level of involvement of industry actors. This has generated concerns, however, about the potential for conflicts between the public interest goals of these bodies and their increasing dependence on the cooperation of profit-motivated actors. This pattern is, of course, familiar at the national level, but its emergence in global standard-setting fora has raised apprehension about the independence of international organizations from the actors they are meant to be regulating (Newell 2003a).

Many see this as part of a broader process of "privatization" of the U.N. system, in which private actors are increasingly carrying out the work of the UN while benefiting from the good reputation of the organization (Utting 2000). This concern reached its height over a proposal by UNDP (United Natives Development Programme) for a Global Sustainable Development Facility, to be funded by leading multinationals that have traditionally drawn fire for the environmental impacts of their operations, including such prominent global players as Shell and Rio Tinto. The development of a Global Compact between the UN and the private sector, to which businesses have been asked to sign up, has generated similar controversy. The Compact lists a series of principles drawn from prominent UN conventions, including the precautionary principle for example, which signatory firms are expected to respect in their business activities. Many have criticized the veneer of legitimacy that association with the UN provides to companies with dubious track records on social, environmental and human rights issues as well as the lack of mechanisms for ensuring that commitments are honored in practice (Corpwatch 2002).

The UNCED (United Nations Conference on Environment and Development) of 1992, also provided a catalyst for a critical "global ecology" literature that focused on business capture of the negotiations leading to the summit (Chatterjee and Finger 1994; Hildyard 1993). This work documented the role of Maurice Strong, former business leader and organizer of the summit, in projecting the achievements of multinational corporations (MNCs) at the expense of discussion about the role of firms in contributing to environmental degradation. Corporate financing of the summit aggravated these claims and evidence of sustained lobbying efforts by businesses and business associations in watering down the Rio agreements added fuel to the chorus of objections to business cooption of the Earth Summit's agenda. MNCs were able to present themselves, alongside governments, as the appropriate stewards of the global commons, bringing their expertise, technology, and capital to the aid of environment. The Business Council for Sustainable Development produced its own *Business Charter for Sustainable Development,* while references in the Rio documents focused on the role of business in finding solutions to environmental problems, rather than the question of their regulation. The involvement of business in delivering responses to environmental threats in this way amounted, in Hildyard's words (1993), to putting the "foxes in charge of the chickens." This episode prompted a spate of studies and critiques questioning the extent to which the greening of business has really taken place (Rowell 1996; Utting 2001). The debate about the appropriate role of firms at such global environmental summits continues in the wake of failed attempts to gain support for a UN Corporate Accountability Convention at the 2002 Johannesburg World Summit on Sustainable Development.

It should be acknowledged, however, that some firms and sectors are far more engaged in the global politics of decision making on environmental issues than others. It is clear, in many instances, that it is larger, multinational firms that are most closely involved in setting standards and not smaller- and medium-sized firms that are less well-organized politically (Dannreuther 2002) and underrepresented in international fora such as the ISO. Indeed, it has been suggested that one of the commercial drivers of private forms of self-regulation, such as ISO 14001 standards, is the desire to keep smaller firms out of profitable markets by raising the barrier to entry and increasing the costs of compliance with standards (Clapp, this volume). Finger (this volume) also shows how water multinationals have been pushing for environmental standards that exceed the capacity of local water providers to meet, setting the stage for their entry into the market for water services.

These contributions help to challenge the idea that businesses are opposed to regulation per se, highlighting the benefits that can accrue to businesses from regulation. Regulatory regimes carry significant implications for competitiveness as costs are imposed unevenly and new market opportunities are created. Regulation can create barriers to entry in a number of ways; regulated industries such as hazardous waste frequently have complex procedures for certifying new processes, thus stabilizing existing technologies and protecting market incumbents. Compliance activities often constitute a relatively fixed cost, thus creating economies of scale (Reinhardt 2000). According to Nash and Ehrenfeld (1997), companies also sometimes initiate private, voluntary mechanisms, such as the U.S. chemical industry's Responsible Care program, to raise public confidence in the industry and discipline poor performers who might attract regulatory pressure for the whole industry.

Some *sectors* of industry are also better organised than others to represent their interests, often reflecting greater experience of long-fought regulatory battles. This would be the case for the chemical and energy sectors, for example. It is also clear that within the coalitions that represent firms as a group of stakeholders, such as the ICC (International Chamber of Commerce), or as a sector on key environmental issues, there are intra-industry battles to define the appropriate position of the organization. Levy shows in his chapter on climate change how the decline of the Global Climate Coalition, for example, was a product of increasingly divergent political and investment strategies among the members of the coalition. The organization of decisionmaking within umbrella federations can have an important bearing, therefore, on which corporate concerns get screened in and out of global debates.

Though there is a long history of examining the lobbying practices of particular industry coalitions (Grant et al. 1998; Greenwood 1997), an emergent theme in the literature on business identifies differences in the lobbying styles and practices of firms, particularly along trans-Atlantic lines. Many have noted the more adversarial style of business lobbying in the United States as opposed to Europe where the approach is focused on dialogue and corridor lobbying. This reflects broader differences in corporate strategy, where firms in the United States have been able to contest the scientific rationales for taking environmental action more openly and directly, where in Europe such positions are untenable. A range of strategic, institutional, and sociocultural factors have been invoked to explain these differences in approach (Levy and Newell 2000; Levy and Kolk 2002). Though differences endure, there has been a notable trend towards convergence between the lobbying styles of business in Europe and North America, as Coen shows in his chapter in this volume. This flows from the increasingly trans-Atlantic nature of capital integration in sectors such as biotechnology, and attempts by global coalitions to construct policy positions that bridge European-U.S. differences. In the case of climate change, Levy and Kolk (2002) argue that initial trans-Atlantic differences in the oil industry's responses were shaped by the home-country institutional environment. Over time, participation in the common global industry and the new issue-specific institutions on climate change has produced some degree of convergence of corporate perspectives, and hence strategic responses.

Globalization and Its Impact on Environmental Governance

Rather than interpret the growing role of business in the international governance of the environment as a trend unique to environmental politics, many see these changes as part of broader shifts in the global political economy associated with globalization. It is argued that these broader shifts in relations between states, markets and civil society need to be understood if we are to make sense of changing patterns of global environmental governance. Changes in the governance and exploitation of environmental resources have been brought about through shifting patterns of production and investment, through changes in the nature of standard-setting and institutional authority, and through increased participation by business and civil society actors in global environmental debates (Newell 1999).

Underlying these trends, many argue, is a shift in the balance of power away from states and towards markets. Debates about the transfer of power from state to market have a long lineage, as our discussion of the literature on bargaining between MNCs and host countries (in chapter 3) illustrates. Yet narratives about globalization have given them a new salience in discussions about the retreat of the state (Strange 1996) in the face of global market pressures. There is debate about whether state power is being reconstituted as "competition states" (Cerny 1990) position themselves to attract increasingly mobile investment, or whether state power may actually be increasing in some parts of the world (Weiss 1998, 2003). The answer to these questions depends very much on which aspect of state power is under scrutiny and the nature of the state in question. Most scholars concur, however, that in the contemporary context business is gaining the upper hand in the state-MNC bargaining process. In the case of the TRIPs agreement, Sell (1999, 172) argues, "In effect, twelve corporations made public law for the world." The growing mobility of capital and economic interdependence of states is widely seen as a key source of increasing MNC leverage (Carnoy et al., 1993). Eden (1993, 47) points to "the importance of multinationals as the engineers or agents of this increasing interdependence." Gill (1995) uses the term "the new constitutionalism" to refer to the processes by which the rights of capital over states are being enshrined in international agreements such as the WTO TRIPs accord, the NAFTA (North American Free Trade Agreement) agreement, and the attempted MAI (Multilateral Agreement on Investment). In the environmental field, the debate within the Convention on Biodiversity (CBD) over access, benefit-sharing and IPRs can be interpreted as a struggle to define corporate entitlements to the commons. Debates about biopiracy and bioprospecting are concerned with the boundaries of corporate access to natural resources and the compensation firms are expected to provide to the communities from which the resources have been extracted. Indeed, it was primarily concerns about how the CBD would affect the access of the U.S. pharmaceutical industry to such resources that led President Bush (senior) to refuse to sign the accord (Rautisala 1997).

These issues are also raised in the literature on business regulation (Drahos and Braithwaite 2000; Picciotto and Mayne 1999), which describes a growing tendency towards *regulation for* business, rather than *regulation of* business (Newell 2001). The difference is between regimes established to facilitate commercial activities, such as market-enabling trade and investment agreements, and those, such as environmental regimes, which are aimed at regulating the effects of the market (market-restricting agreements) (Egan and Levy 2001). Evidence of an imbalance towards the protection of corporate rights (regulation for), rather than the promotion of corresponding corporate responsibilities (regulation of), is to be found in the contrast between recurrent failures to agree binding standards on the international conduct of MNCs and the rise of agreements whose aim is to allocate and secure property rights for firms and facilitate market access.

Once again, the increasing power of MNCs should not be equated with a zero-sum loss of state power. Indeed, it is the state's resources to which MNCs wish to secure access, and it is the very division of the world into competing national states that provides capital with its structural power (Gill and Law 1993). It is states that sign the market-enabling accords strengthening the bargaining power of business and removing constraints on capital mobility (Helleiner 1994). Firms also need states, albeit pliable ones, just as states need firms to secure their growth and employment goals. Facilitating new markets, whether for electricity or tradable emissions permits, requires a weighty regulatory and administrative structure, leading to the apparent paradox of simultaneous deregulation and reregulation. This can obviously create tensions when the state is expected to act simultaneously as a promoter and regulator of an industry (Jasanoff 1995); in some of the cases in this volume, these tensions play out across governmental agencies and departments within the same state.

The extent to which these trends towards the erosion or reconfiguration of state power apply also depends very much on the state in question. Reno (1996) and Frynas (1998) show that in a context of weakened state structures, the power and privileges of business groups are enhanced such that "At the very time statist theorists were 'bringing the state back in,' the importance of private actors in formulating state policies was actually increasing in Africa and elsewhere in the developing world" (Cox 1996, 5). Just as theories of the firm seek to get inside the "black box" of the firm (Amoore 2000), so too theories of the state have sought to look inside the state for an understanding of the importance of bureaucratic politics and policy networks. As was argued above, this is significant because businesses are better connected with some parts of the state than others and these patterns of influence extend into the international sphere (Newell 2003b). Studies of international governance need to develop, therefore, a much more contextualized understanding of the role of the state and intrastate politics, which both refract and define the nature of global engagements.

Partly because of the perceived collusion between state and business actors, heightened in a context of globalization, environmental activists have sought to contest the power of business in new ways. Increasingly, activists target businesses directly, rather than channel their lobbying through the state, reflecting the shift in the balance of power between states and markets. This strategic shift in the campaigning activities of environmental NGOs has taken a number of forms, from confrontational approaches to more collaborative strategies of engagement. There is a growing literature that seeks to analyse the significance of "partnerships" between business and NGOs in the context of broader debates about corporate social responsibility. Some work focuses on examining the conditions in which firms and NGOs can come together to construct "win-win" solutions to environmental problems (Long and Arnold 1995). Other work looks at whether the negotiation of codes of conduct and other forms of soft regulation constitute forms of civil regulation; civil-society-based regulation that helps to plug some of the governance gaps left by governments' own inability or unwillingness to regulate the environmental impact of businesses' activities (Bendell 2000; Newell 2001). Alongside this, there has also been increasing interest in the role of activist groups and social movements in challenging corporate hegemony (Cromwell 2001; Korten 1995; Karliner 1997; Mittelman 1998; Newell 2002).

Increasing emphasis on the civil society's own relations with business represents a welcome, if currently underdeveloped, contribution to reflections about the role of business in global governance. Earlier work by Cox (1996) was more narrowly focussed on how "The power of business is articulated through the interaction of three entities within the international system; the nation-state, the multinational corporation and domestic firms" (1996, 1). While such work usefully drew attention to the importance of the sectors in which firms are based as a determinant of their policy preferences and explicitly sought to develop a theory of business conflict, its application to global environmental politics is limited by its neglect of the role of civil society actors.

These disparate bodies of work point to the changing nature of environmental governance, resulting, at least in part, from broader shifts in the global political economy. Each area of work captures different dimensions and manifestations of this trend, from the privatization of the UN and the construction of private regimes to reconfigured roles for state, business, and civil society actors as each adapts to the changing power of the other. Accounting for these changes theoretically clearly requires us to think more imaginatively across levels of analysis and across disciplines and serves to highlight the acute limitations of traditional patterns of theorizing.

Conclusion

While orthodox forms of theorising about global environmental politics underestimate the importance of business actors in processes of regime formation and to the overall effectiveness of international environmental policy, there are strands of thinking that provide multiple entry points for constructing a new account of the role of the firm in global environmental governance. These include the work on private regimes, the long history of debates about regulating the activities of multinational companies, contemporary concerns with the privatization of the UN and corporate penetration of global policy processes, as well as the ways in which civil society actors have sought to contest this concentration of power. Our tools for thinking about the links between the political and investment strategies of firms on environmental issues and the content and direction of international environmental regimes remain underdeveloped, however.

By opening up the "black box" of the firm to more critical scrutiny, we may be better placed to locate the linkages and connections between inter- and intrafirm decision making and the activities of firms in international environmental fora. Amoore's (2000, 183) notion of the contested firm may provide a useful starting point in this regard, as it seeks to go beyond treating firms as "actors, reactors and transmitters of global imperatives." It requires us to do more than regard firms as transmission belts between national and global levels of analysis and to challenge their predominant construction in IPE as atomized, rational, unitary actors, an approach that is subject to many of the failings of conventional IR theory. As Amoore (2000, 185) notes "Put simply, orthodox understandings of the firm in IPE tend neither to open up the firm to examine the social power relations within, nor to look outside at their extension into wider social contests." We have then a classic case of mutual neglect, where thinking about corporate strategy within organization and management theory has failed to make connections to global processes of governance that increasingly impact on the strategies of leading firms, and debates about international environmental governance have overlooked insights from these disciplines that shed light on emerging enquiries into the role of the firm in global governance. The next chapter shows how an approach drawing on the work of Gramsci, among others, is useful in this regard.

The limitations of many of the approaches described above underscores the need to develop a political economy approach, one that looks at the ways in which environmental regimes, broadly construed, are embedded within the global political economy, at once reflecting and advancing the global economic forces of which they are a product. The location of states in a dynamic global capitalist economy, faced by the economic discipline of mobile capital and financial markets, may severely constrain the range of policy options considered viable for tackling environmental problems. The degree to which different actors are influential in such deal-brokering, and therefore an understanding of why some outcomes prevail over others, requires an understanding of the different dimensions of power not just within a regime, but in multiple political, economic and social sites within the global political economy. Placing relations between states, markets, and civil society centrally, informs our understanding of why some policy solutions are privileged over others, why, for instance, market-based solutions are promoted over more "command-and-control interventions," why some sectors are protected from regulation more than others, and why some policy options are kept off the agenda altogether. In the next chapter we attempt to develop such a political economy approach to bring some theoretical coherence and insight to these processes.

Note

1. Exceptions include Newell (2000a).

References

Amoore, L. (2000). International political economy and the contested firm. *New Political Economy*, 5(2), 183–204.

Bachrach, P., & Baratz, M. S. (1963). Two faces of power. *American Political Science Review*, 56, 947–952.

Bendell, J. (2000). Civil regulation: A new form of democratic governance for the global economy? In J. Bendell (Ed.), *Terms for endearment: Business, NGOs and sustainable development* (pp. 239–255). Sheffield, UK: Greenleaf.

Benedick, R. E. (1991). *Ozone diplomacy: New directions in safeguarding the planet.* Cambridge, MA: Harvard University Press.

Boehmer-Christiansen, S. (1994). Global climate policy: The limits of scientific advice. Part 1. *Global Environmental Change, 4*(2), 140–159.

Caldwell, L. K. (1988). Beyond environmental diplomacy: The changing institutional structure of international cooperation. In J. Carroll (Ed.), *International environmental diplomacy* (pp. 13–28). Cambridge, MA: Cambridge University Press.

Carnoy, M., Castells, M., & Cohen, S. (Eds.) (1993). *The new global economy in the informational age.* University Park: Pennsylvania State University Press.

Cerny, P. (1990). *The changing architecture of politics: Structure, agency and the future of the state.* London: Sage Publications.

Chatterjee, P., & Finger, M. (1994). *The Earth brokers: Power, politics and world development.* London: Routledge.

Cohen, R., & Rai, S. (Eds.) (2000). *Global social movements.* London: Athlone Press.

CorpWatch. (2002). *Greenwash + 10: The UN's global compact, corporate accountability, and the Johannesburg Earth Summit.* California: CorpWatch/Tides Center.

Cox, R. W. (Ed.) (1996). *Business and the state in international relations.* Boulder, CO: Westview.

Crenson, M. (1971). *The Unpolitics of air pollution.* Baltimore: John Hopkins Press.

Cromwell, D. (2001). *Private planet: Corporate plunder and the fight back.* Oxon, UK: Jon Carpenter.

Cutler, C., Haufler, V., & Porter, T. (Eds.) (1999). *Private authority and international affairs.* Albany, NY: State University of New York Press.

Dannreuther, C. (2002). Globalisation and SME's: The European experience. In P. Newell, S. Rai, & A. Scott (Eds.), *Development and the challenge of globalisation* (pp. 115–127). London: ITDG Press.

Drahos, P., & Braithwaite, J. (2000). *Global business regulation.* Cambridge: Cambridge University Press.

Eden, L. (1991). Bringing the firm back in: Multinationals in IPE. *Millennium, 20*(2), 197–224.

Eden, L. (1993). Bringing the firm back in: Multinationals in the international political economy. In L. Eden & E. Potter (Eds.), *Multinationals in the global political economy* (pp. 25–58). New York: St. Martin's Press.

Edwards, M., & Gaventa, J. (2001). *Global citizen action.* Boulder, CO: Lynne Rienner.

Egan, D., & Levy, D. (2001). International environmental politics and the internationalization of the state: The cases of climate change and the MAI. In D.

Stevis & V. Assetto (Eds.), *The international political economy of the environment: Critical perspectives* (pp. 63–85). Boulder, CO: Lynne Rienner.

Falkner, R. (2003). Private environmental governance and international relations: Exploring the links. *Global Environmental Politics, 3*(2), 72–88.

Finger, M., & Tamiotti, L. (1999). The emerging linkage between the WTO and the ISO: Implications for developing countries. In *IDS Bulletin*, Vol. 30 No. 3, July. Brighton: IDS.

Frynas, G. (1998). Political instability and business: Focus on Shell in Nigeria. *Third World Quarterly, 19*(3), 457–478.

Gale, F. (1998). "Cave, Cave! Hic dragones": A neo-Gramscian deconstruction and reconstruction of international regime theory. *Review of International Political Economy, 5*(2), 252–284.

Garcia-Johnson, R. (2000). *Exporting environmentalism*. Cambridge, MA: MIT Press.

Gill, S. (1995). Theorising the interregnum: The double movement and global politics in the 1990s. In B. Hettne (Ed.), *International political economy: Understanding global disorder*. (pp. 45–99). London: Zed Books.

Gill, S., & Law, D. (1993). Global hegemony and the structural power of capital. In S. Gill (Ed.), *Gramsci, historical materialism and international relations* (pp. 93–124). Cambridge, UK: Cambridge University Press.

Grant, W., Matthews, D., & Newell, P. (2000). *The effectiveness of EU environmental policy*. Basingstoke, UK: Macmillan.

Grant, W., Paterson, W. E., & Whitson, C. (1988). *Government and the chemical industry*. Oxford: Clarendon Press.

Greenwood, J. (1997). *Representing interests in the European Union*. London: Macmillan.

Haas, P. (1990). Obtaining international environmental protection through epistemic consensus. *Millennium, 19*(3), 347–363.

Haas, P., Keohane, R., & Levy, M. (1993). *Institutions for the Earth: Sources of effective environmental protection*. Cambridge, MA: MIT Press.

Hajer, M. (1997). *The politics of environmental discourse: Ecological modernization and the policy process*. Oxford, UK: Oxford University Press.

Hasenclever, A., Mayer, P., & Rittberger, V. (1997). *Theories of international regimes*. Cambridge, UK: Cambridge University Press.

Haufler, V. (1995). Crossing the boundary between public and private. In V. Rittberger (Ed.), *Regime theory and international relations* (pp. 94–112). Oxford: Clarendon Press.

Helleiner, E. (1994). From Bretton Woods to global finance: A world turned upside down. In R. Stubbs & G. Underhill (Eds.), *Political economy and the changing global order* (pp. 163–175). Basingstoke, UK: Macmillan.

Helleiner, E. (1996). International political economy and the Greens. *New Political Economy*, *1*(1), 59–77.

Higgott, R., Underhill, G., & Bieler, A. (Eds.) (1999). *Non-state actors and authority in the global system*. London: Routledge.

Hildyard, N. (1993). Foxes in charge of the chickens. In W. Sachs (Ed.), *Global ecology* (pp. 22–35). London: Zed Books.

Jakobsen, S. F. (1997). Transnational NGO activity, international opinion, and science—Crucial dynamics of developing country policymaking on climate. Paper for conference on *Non-state actors and authority in the global system*, Warwick University, October 31st–November 1st.

Jasanoff, S. (1995). Product process or programme: Three cultures and the regulation of biotechnology. In M. Bauer (Ed.), *Resistance to new technology* (pp. 311–334). Cambridge, UK: Cambridge University Press/The National Museum of Science and Industry.

Karliner, J. (1997). *The corporate planet: Ecology and politics in an age of globalisation*. San Francisco: Sierra Club Books.

Keeley, J. (1990). Toward a Foucauldian analysis of international regimes. *International Organisation*, *44*(1), 83–105.

Keohane, R. (1984). *After hegemony: Cooperation and discord in the world political economy*. Princeton, NJ: Princeton University Press.

Keohane, R. O., & Milner, C. (1997). *Internationalization and domestic politics*. Cambridge, UK: Cambridge University Press.

Keohane, R., & Nye, J. (1972). *Transnational relations and world politics*. Cambridge, MA: Harvard University Press.

Korten, D. (1995). *When corporations rule the world*. London: Earthscan.

Krasner, S. D. (Ed.) (1983). *International regimes*. Ithaca, NY: Cornell University Press.

Krut, R., & Gleckman, H. (1998). *ISO 14001: A missed opportunity*. London: Earthscan.

Laferrière, E. (1996). Emancipating international relations theory: An ecological perspective. *Millennium*, *25*(1), 53–75.

Laferrière, E., & Stoett, P. (1999). *International relations theory and ecological thought: Towards a synthesis*. London: Routledge.

Lee, K., Humphreys, D., & Pugh, M. (1998). Privatisation in the United Nations system: Patterns of influence in three intergovernmental organisations. *Global Society*, *11*(3), 339–359.

Levy, D. L., & Kolk, A. (2002). Strategic responses to global climate change: Conflicting pressures on multinationals in the oil industry. *Business and Politics*, *4*(3), 275–300.

Levy, D. L., & Newell, P. (2000). Oceans apart? Business responses to global environmental issues in Europe and the United States. *Environment*. *42*(9), 8–20.

Lipsky, M. (1980). *Street-level bureaucracy: Dilemmas of the individual in public services*. New York: Russell Sage Foundation.

Litfin, K. (1994). *Ozone discourses*. New York: Columbia University Press.

Long, F. J., & Arnold, J. (1995). *The power of environmental partnerships:* Management Institute for Environment and Business. Orlando, FL: Dryden Press/ Harcourt Brace and College.

Mintzer, I. M., & Leonard, J. A. (Eds.) (1994). *Negotiating climate change*. Cambridge, UK: Cambridge University Press.

Mittelman, J. (1998). Globalisation and environmental resistance politics. *Third World Quarterly, 19*(5), 847–872.

Murphy, D., & Bendell, J. (1997). *In the company of partners*. Bristol: Policy Press.

Nash, J., & Ehrenfeld, J. (1997). Codes of environmental management practice: Assessing their potential as a tool for change. *Annual Review of Energy and Environment, 22*, 487–535.

Newell, P. (1999). Globalisation and the environment: Exploring the connections. *IDS Bulletin, 30*(3), 1–8.

Newell, P. (2000a). *Climate for change: Non-state actors and the global politics of the greenhouse*. Cambridge, UK: Cambridge University Press.

Newell, P. (2000b). Environmental NGOs and Globalization. In R. Cohen & S. Rai (Eds.), *Global social movements*. London: Athlone Press.

Newell, P. (2001). Managing multinationals: The governance of investment for the environment. *Journal of International Development, 13*, 907–919.

Newell, P. (2002). From responsibility to citizenship: Corporate accountability for development. *IDS Bulletin, 33*(2) *Making rights real: Exploring citizenship, participation and accountability*. Brighton, UK: IDS.

Newell, P. (2003a). Globalisation and the governance of biotechnology. *Global Environmental Politics, 3*(2), 56–72.

Newell, P. (2003b). Biotech firms, biotech politics: Negotiating GMOs in India. *IDS Working Paper* 201, September. Brighton: IDS.

Newell, P., & Mackenzie, R. (2000). The Cartagena protocol on biosafety: Legal and political dimensions. *Global Environmental Change, 10*, 313–317.

Nowell, G. P. (1996). International relations theories: Approaches to business and the state. In R. W. Cox (Ed.), *Business and the state in international relations* (pp. 181–197). Boulder, CO: Westview.

Nye, J., & Keohane, R. (1972). Transnational relations and world politics. In R. Keohane & J. Nye (Eds.), *Transnational relations and world politics* (pp. ix–xxiv). Cambridge MA: Harvard University Press.

O'Brien, R., Goetz, A. M., Scholte, J. A., & Williams, M. (2000). *Contesting global governance*. Cambridge, UK: Cambridge University Press.

Oye, K. A., & Maxwell, J. (1994). Self-interest and environmental management. *Journal of Theoretical Politics, 64,* 593–624.

Paterson, M. (2001). *Understanding global environmental politics: Domination, accumulation and resistance.* Basingstoke, UK: Palgrave.

Picciotto, S., & Mayne, R. (1999). *Regulating international business: Beyond liberalization.* Basingstoke, UK: Macmillan.

Putnam, R. (1988). Diplomacy and domestic politics: The logic of two-level games. *International Organisation, 42*(3), 427–460.

Rautisala, K. (1997). The domestic politics of global biodiversity protection in the United Kingdom and United States. In M. Schreurs & E. Economy (Eds.), *The internationalisation of environmental protection* (pp. 42–74). Cambridge, Cambridge University Press.

Reinhardt, F. L. (2000). *Down to earth: Applying business principles to environmental management.* Boston, MA: Harvard Business School Press.

Reno, W. (1996). Business conflict and the shadow state: The case of West Africa. In R. W. Cox (Ed.), *Business and the state in international relations* (pp. 150–151). Boulder, CO: Westview.

Risse-Kappen, T. (1995). *Bringing transnational relations back In.* Cambridge, UK: Cambridge University Press.

Rosenau, J., & Czempiel, E. O. (Eds.) (1992). *Governance without government.* Cambridge, UK: Cambridge University Press.

Rowell, A. (1996). *The green backlash.* London: Routledge.

Saurin, J. (1996). International relations, social ecology and the globalisation of environmental change. In J. Vogler & M. Imber (Eds.), *The environment and international relations* (pp. 77–99). London: Routledge.

Sell, S. K. (1999). Multinational corporations as agents of change: The globalization of intellectual property rights. In C. Cutler, V. Haufler, & T. Porter, *Private authority and international affairs.* Albany: State University of New York Press.

Sell, S. K. (2003). *Private power, public law: The globalization of intellectual property rights.* Cambridge, UK: Cambridge University Press.

Stevis, D., & Assetto, V. (Eds.) (2001). *The international political economy of the environment: Critical perspectives.* Boulder, CO: Lynne Rienner.

Stopford, J., & Strange, S. (Eds.) (1991). *Rival states, rival firms: Competition for world market shares.* Cambridge, UK: Cambridge University Press.

Strange, S. (1983). Cave! hic dragones: A critique of regime analysis. In S. Krasner (Ed.), *International regimes* (pp. 337–354). Ithaca: Cornell University Press.

Strange, S. (1994a). *States and markets* (2nd ed.), London: Pinter.

Strange, S. (1994b). Rethinking structural change in the International Political Economy: States, firms and diplomacy. In R. Stubbs & G. Underhill (Eds.),

Political economy and the changing global order (pp. 103–116). Basingstoke: Palgrave Macmillan.

Strange, S. (1996). *The retreat of the state.* Cambridge, UK: Cambridge University Press.

Stubbs, R., & Underhill, G. (Eds.) (1994). Political economy and the changing global order. Basingstoke: Palgrave Macmillan.

Utting, P. (2000). UN-Business partnerships: Whose agenda counts? Paper presented at seminar on Partnerships for Development or Privatization of the Multilateral System, North-South Coalition, Oslo, Norway, December 8, 2000.

Utting, P. (Ed.) (2001). *The greening of business in developing countries: Rhetoric, reality and prospects.* London: Zed Books.

Vogler, J. (1995). *The global commons: A regime analysis.* Sussex: Wiley.

Wapner, P. (1996). *Environmental activism and world civic politics.* Albany, NY: State University of New York Press.

Ward, H. (1996). Game theory and the politics of global warming: The state of play and beyond. *Political Studies, XLIV,* 850–871.

Weiss, L. (1998). *The myth of the powerless state.* Ithaca: Cornell University Press.

Weiss, L. (2003). *States in the global economy: Bringing domestic institutions back in.* Cambridge, UK: Cambridge University Press.

Williams, M. (1996). International political economy and global environmental change. In J. Vogler & M. Imber (Eds.), *The environment and international relations* (pp. 41–58). London: Routledge.

Wynne, B. (1994). Scientific knowledge and the global environment. In M. Redclift & T. Benton (Eds.), *Social theory and the global environment* (pp. 169–190). London: Routledge.

Young, O. (1989). *International cooperation: Building regimes for natural resources and the environment.* Ithaca, NY: Cornell University Press.

Young, O. (1998). *Global governance: Learning lessons from the environmental experience.* Cambridge, MA: MIT Press.

Young, O., & Von Molkte, K. (1994). The consequences of international environmental regimes. *International Environmental Affairs, 6*(4), 348–368.

3

A Neo-Gramscian Approach to Business in International Environmental Politics: An Interdisciplinary, Multilevel Framework[1]

David L. Levy and Peter J. Newell

Introduction

Business plays a central role in the industrial activities that account for many of the adverse environmental impacts afflicting the planet; at the same time, its technological and financial resources could potentially make a major contribution toward mitigating and reversing these impacts. Business also is a key political actor in negotiating, structuring, and implementing environmental policy at the national and international levels. Nevertheless, theoretical tools for understanding the relationship of business to international environmental governance are underdeveloped. There has been little scholarly attention to the reasons why business supports (or opposes) some environmental regimes more than others. We need a better understanding of the connections between corporate strategies in the market and political spheres, and of the underlying processes by which corporate perceptions of interests develop. The relationships need to be explored between the emergence of international regimes of environmental governance and the growth of corporate environmental management practices, private-public partnerships, and private environmental codes and standards.

A focus on the role of the private sector in environmental governance suggests the need for a political economy approach. The interdisciplinary framework developed here bridges macrolevels and microlevels of analysis by bringing together perspectives from International Relations (IR) with theories of management and organization. In broad terms, we view the uneven and fragmented nature of international environmental governance as the outcome of a process of bargaining, compromise, and alliance formation at the level of specific regimes. These negotiations, which are simultaneously shaped by and constitutive of the broader structures of global governance, engage a range of actors including states and transnational organizations, businesses and industry associations, and social forces such as environmental and labor groups. Organizations such as the UN and the World Trade Organization (WTO) serve in multiple capacities; as fora for bargaining, as targets of policy, and as semi-autonomous agents in their own right.

The development of each environmental regime is shaped by micro-processes of bottom-up bargaining and constrained by existing macrostructures of production relations and ideological formations. These structures, which themselves are the outcome of historical conflicts and compromises, ensure that the bargaining process is not a pluralistic contest among equals, but rather is embedded within broader relations of power. Nevertheless, the complexity and dynamic nature of the bargaining process, within which alliances, interests, and capacities of actors can shift and mutate, lends a degree of indeterminacy to the compromises reached over the form and mechanisms of individual regimes. Sensitivity to a strategic dimension of power suggests that intelligent agency can sometimes outmaneuver resource-rich adversaries.

Beyond Regime Theory

Despite the extensive development of regime theory in the IR literature, the approach has a number of limitations, as discussed in the previous chapter. It has primarily been used to explain international mechanisms for "resolving conflicts, facilitating cooperation, or more generally, alleviating collective-action problems in a world of interdependent actors" (Young 1994, 3). This functionalist orientation presumes that regimes are benign entities negotiated in a pluralistic context to provide public goods such as environmental protection. Moreover, regime theory has been much criticized for its state centricity and neglect of nonstate actors. The neo-Gramscian framework developed in this chapter, by contrast, proposes that international environmental governance is a profoundly political process in which business, NGOs, and state agencies engage in contests over the structures and processes that constrain and order industrial activities giving rise to environmental impacts. This approach suggests a broad notion of environmental governance, one that encompasses the industrial structures and international financial and trade institutions that provide pattern and structure to industry competitive dynamics and corporate practices. This opens up analytical space to examine the embeddedness of particular environmental regimes within broader economic and political structures of the global economy, and the linkages between domestic and international politics (DeSombre 2000; Schreurs 1997). The framework also highlights the political nature of strategies to protect market position, legitimacy, and autonomy in the face of environmental issues; technological innovation, partnerships with NGOs, and the development of private standards can all be viewed as political elements of environmental governance systems, in this broader sense. The focus on strategy also puts emphasis on the processes of political contestation and compromise as actors attempt to build alliances and frame public debates over science and economics in particular ways.

Gramscian ideas provide a conceptual linkage between corporate strategy and international relations in constructing a political economy of international environmental governance.[2] Our framework offers a number of unique insights. It addresses relationships between national and international levels of analysis, between states and nonstate actors, and between agency and structural relations of power. It points to particular patterns of strategies likely to be adopted in bargaining over complex regimes, and highlights the dynamic, and somewhat indeterminate path of regime evolution. Finally, it suggests a strategic concept of power that presents opportunities for resource-poor groups to outmaneuver rivals.

Gramsci's Politics

Perhaps Gramsci's most significant contribution to political thought is the concept of hegemony in explaining social order. Hegemony is not dependent on coercive control by a small elite, but rather rests on coalitions and compromises that provide a measure of political and material accommodation with other groups, and on ideologies that convey a mutuality of interests. Hegemonic stability is rooted in the institutions of civil society, such as the church, academia, and the media, which play a central role in ideological reproduction, providing legitimacy through the assertion of moral and intellectual leadership and the projection of a particular set of interests as the general interest. Civil society, in Gramsci's view, has a dual existence. As the ideological arena in which hegemony is secured, it represents part of the "extended state," complementing the coercive potential of state agencies. However, the relative autonomy of civil society from economic structures and from state authority turns the ideological realm into a key site of political contestation.

An "historical bloc," in Gramscian terms, exercises hegemony through the coercive and bureaucratic authority of the state, dominance in the economic realm, and the consensual legitimacy of civil society. Gramsci used the term "historical bloc" to refer to the alliances among various social groupings and also to the specific alignment of material, organizational, and discursive formations that stabilize and reproduce relations of production and meaning. These two meanings of "historical bloc" are closely related, for the ability to mobilize an effective alliance requires not just economic side-payments but also discursive frameworks that actively constitute perceptions of interests. For Gramsci, hegemony entails:

not only a unison of economic and political aims, but also intellectual and moral unity . . . the development and expansion of the [dominant] group are conceived of, and presented, as being the motor force of a universal expansion. . . . In other words, the dominant group is coordinated concretely with the general interests of the subordinate groups. (Gramsci 1971, 181)

Hegemony is also contingent and unstable. The economic and ideational realms evolve in dialectical tension, generating underlying fault lines and contradictions. Gramsci was acutely sensitive to the resulting dynamics: "What is this effective reality? Is it something static and immobile, or is it not rather a relation of forces in continuous motion and shift of equilibrium?" (Gramsci 1971, 172). This understanding of the complex dynamic nature of social systems led Gramsci to emphasize the importance of agency and strategy in challenging groups with superior resources. Drawing from Machiavelli, Gramsci posited that the political party could serve as the "Modern Prince," who could analyze the "relations of forces" to reveal weaknesses and points of leverage, and possess the organizational capacity to intervene during critical windows of opportunity. Gramsci (1971, 172) warned against fatalism that stems from overly deterministic, structural accounts of history, and also against utopianism that results from excessive faith in unconstrained agency: "The active politician is a creator, an initiator; but he neither creates from nothing nor does he move in the turbid void of his own desires and dreams. He bases himself on effective reality . . . but does so in order to dominate and transcend it." In other words, social change requires incremental measures that take into account the constraints of existing structures, while developing strategies for more radical change that alter those structures.

Gramsci outlined two particular forms of strategy commonly evinced in social conflicts. The term "passive revolution" was used to describe a process of reformist change from above, which entailed extensive concessions by relatively weak hegemonic groups, often in the guise of populist or nationalist programs, in an effort to preserve the essential aspects of social structure. The concept of "war of position" employed a military metaphor to suggest how subordinate groups might avoid a futile frontal assault against entrenched adversaries; rather, the war of position constitutes a longer term strategy, coordinated across multiple bases of power, to gain influence in the cultural institutions of civil society, develop organizational capacity, and to win new allies. As in a game of chess, power lies not just in the playing pieces, but in the configuration of forces relative to each other and to adversaries, and each set of moves and countermoves opens up new fissures and presents fresh possibilities to prise open the seams of a historical bloc. Successful strategy thus requires careful analysis: "It is the problem of the relations between structure and superstructure which must be accurately posed if the forces which are active in the history of a particular period are to be correctly analyzed and the relations between them determined" (Gramsci 1971, 177).

Gramsci and International Relations

If the role of business has been somewhat neglected in regime theory, it has taken center stage in the work of a number of scholars who have applied Gramsci to questions of international relations. Cox (1987, 357) argues that this approach "regards class formation and the formation of historic blocs as the crucial factor in the transformation of global political and social order," generating a bottom-up understanding of the world economy and state system that avoids the economic determinism of world systems theory. Cox and others describe the ascendancy of a transnational historical bloc comprising a managerial elite from MNCs, professionals from NGOs and academia, and governmental agencies (Murphy 1998; Robinson 1996). Cox contends that we are witnessing the growth and coordination at a global level of economic structures, neoliberal and consumerist ideologies, and a set of political institutions such as the WTO and International Monetary Fund. Sklair points to the strategic function of transnational industry groupings such as the Trans-Atlantic Business Dialogue and the European Roundtable of Industrialists in creating the infrastructure of the emerging bloc (Sklair 1997). At the center of this bloc, Cox (1987) argues, is a transnational managerial class, which, despite internal rivalries, displays an "awareness of a common concern to maintain the system":

Various institutions have performed the function of articulating strategies in this common concern: the Trilateral Commission, the OECD, the IMF, and the World Bank all serve as foci for generating the policy consensus for the maintenance and defense of the system . . . Prestigious business schools and international management training programs socialize new entrants to the values, lifestyles, language (in the sense of shared concepts, usages, and symbols), and business practices of the class. (p. 359)

Gill refers to the dominant ideology of the transnational elite as "disciplinary neoliberalism," which incorporates a faith in market forces, privatization, unfettered international trade and investment, and minimal provision of social services (Gill 1995). He describes the surveillance mechanisms that impose discipline on states, companies, and individuals in the new order, from the monitoring of inflation rates and budget deficits to corporate and personal credit ratings. The result of this restructuring is a "new constitutionalism" in which the rights of capital over states are enshrined in global accords.

The application of Gramscian thought to current trends in the international political economy has not been without critique. Germain and Kenny (1998) have questioned whether Gramsci indeed offers a coherent perspective on the relationship between economic structure, ideology, and agency. As Rupert acknowledges in a response, "Gramsci's legacy is fragmentary, fraught with analytical and political tensions, and eminently contestable" (Rupert 1998, 427). This is hardly surprising, given the unfinished nature of Gramsci's notes and the complexity of the theoretical challenge. Gramsci's value lies, rather, in the inspiration he has given to contemporary theorists in their sophisticated treatments of these issues (Giddens 1984; Hall 1986).

Germain and Kenny's critique of neo-Gramscian IR for neglecting processes of resistance is pertinent for some IR scholars who provide a rather deterministic reading of Gramsci. The writings of Cox, Robinson, and van der Pijl, for example, appear in places to reflect a top-down, overly economistic depiction of structures of governance, which positions the national state as a servant of international capital. Robinson argues that national states are converted into "transmission belts and filtering devices for the imposition of the transnational agenda" (Robinson 1996, 19). This stream of work also suffers from a neglect of the role of ideology. Though van der Pijl (1998, 114) notes the role of the American Council on Foreign Relations and the Royal Institute for International Affairs in attempting to forge a trans-Atlantic policy consensus, his description (pp. 110–112) of the rapid ideological shift in the early 1930s among policy elites from liberalism to state monopolism and protectionism suggests an opportunistic response that belies Gramsci's conception of ideology "as a material force" (1971, 165). The writing of these neo-Gramscians is closer in spirit to the business conflict model, which attempts to explain foreign policy in terms of competition among industrial blocs (Cox 1996; Skidmore-Hess 1996). Gill's work, by contrast, is more attuned to the significance of discursive formations and the opportunities for contestation (Gill 1995).

Germain and Kenny also question the relevance of Gramsci's analysis of the state-civil society relationship to contemporary international relations, particularly the meaning of international civil society and hegemony in the absence of a supranational state, or a "corresponding structure of concrete political authority" (Germain and Kenny 1998). As Rupert and Murphy emphasize in their responses, the problem is not one of trying to discern the intent or truth of Gramsci's original text, but rather whether the core concepts retain value (Murphy 1998; Rupert 1998). We argue that hegemony retains validity in describing the specific ensemble of economic and discursive relations that bind a network of actors within a framework of international institutions. This framework includes international agencies that exercise normative and disciplinary sanctions, if not sovereign powers.

Similarly, Gramsci's concept of civil society has application if emergent international NGOs play the same dual role envisaged by Gramsci; as semi-autonomous arenas of cultural and ideological struggle, and also as key allies in securing hegemonic stability. Environmental NGOs such as World Resources Institute and Environmental Defense have become major advocates for market solutions and private partnerships, providing legitimacy for market-based approaches to environmental problems (Helvarg 1994; Rowell 1996). Boehmer-Christensen's work (1996) also shows that international scientific groups such as the IPCC are more embedded in the political process than suggested by conventional accounts of epistemic communities. At the very least, a Gramscian perspective should provoke a more critical engagement with pluralist accounts of global civil society that champion NGOs as the autonomous social groups balancing the power of states and capital (Lipschutz 1992; Wapner 1995).

Despite Gramsci's focus on hegemony in the national context, he did recognize that capitalism and class consciousness traversed national boundaries. Some of his work was comparative, examining the specific historical configurations of economic, ideological, and political forces in different countries, but he also addressed the shifting relationships among major states, such as the growth of regionalism (Gramsci 1995, 191–269, especially 232–233). He also noted the emergence of "international public and private organizations that might be the shapeless and chaotic civil society of a larger, economically concrete social order, and that certainly promoted such an order" (Murphy 1998, 423). Gramsci's analysis was acutely sensitive to the interplay of forces operating at these multiple levels. His contribution to our understanding of environmental governance, therefore, lies less in his scattered notes on international politics and economics, and more in the concept of hegemonic formations as complex dynamic systems comprising overlapping and interpenetrating subsystems:

> international relations intertwine with these internal relations of nation-states, creating new, unique and historically concrete combinations. A particular ideology, for instance, born in a highly developed country, is disseminated in less developed countries, impinging on the local interplay of combinations. This relation between international forces and national forces is further complicated by the existence within every State of several structurally diverse territorial sectors, with diverse relations of force at all levels (1971, 182).

Gramsci, Organization Theory, and Corporate Strategy

The IR literature tends to treat corporate interests at an abstract, aggregate level; capital rather than corporations. A political economy approach, while recognizing the embeddedness of regimes in broader structures, needs to address the specific conditions under which firms engage with particular issue arenas; a theory of the firm as a political actor is needed. Management and organization theory offers several perspectives on corporate political strategy and environmental management, but these tend to be somewhat disconnected from issues of political economy and international governance. Later, we will examine how a neo-Gramscian approach can enrich these managerialist perspectives.

The extant corporate political strategy (CPS) literature is primarily concerned with empirical investigation and categorization of the drivers and forms of CPS (Getz 1997). Early writing in the field emphasized corporate dependence on government policy and characterized strategies along a continuum, from reactive to more effective, proactive approaches (Mahon 1983; Weidenbaum 1981). While much of the literature has viewed CPS as a set of nonmarket activities quite distinct from market-oriented strategies, Baron has argued for their integration and Schuler has noted that political strategies frequently serve as a substitute for failing competitive strategies (Baron 1997; Schuler 1996).

The CPS literature draws from a disparate set of conceptual frameworks. Political strategy at the industry level has been viewed as a form of collective action; the question is then one of the costs and benefits of participation. This perspective suggests that industries are more likely to undertake coordinated action when firms face a common threat, when large economies of scale from cooperation are available, and when industry concentration enables a few large firms to bear the costs (Lehne 1993). Another stream of research examines the strategic use of regulation by firms to increase costs for competitors or reduce the threat of competitive market entry (Leone 1986). Oster found that firms pursue political strategies that hurt rivals even when the outcome was detrimental to the industry as a whole (Oster 1982). Shaffer's study of the U.S. automobile industry's response to CAFE standards points to the dynamic evolution of political strategies as regulations evolve and confront companies with differential compliance costs (Shaffer 1992).

Several contributions have examined firm-level and institutional variables that affect the political strategy formulation process. Boddewyn and Brewer (1994) have asserted that the intensity of political behavior is likely to be greater when the stakes are higher, opportunities for leverage are greater, and firms' political competencies are more developed. Moreover, this political behavior is likely to be conflictual rather than accommodating when potential policies have a high strategic salience, when the situation is perceived as zero-sum, and firms have sufficient power to affect the outcome. Writing a decade before climate change became an issue for the fossil fuel industry, Gladwin and Walter (1980) argued that secure supplies and stable demand are the "jugular veins" of MNCs in the oil industry, such that any threat would likely provoke an assertive and uncooperative corporate response. Hillman and Hitt (1999) observed that CPS varies across countries, depending on the political context. In corporatist nations, for example, firms are likely to use relational approaches to CPS, in which the firm builds a broad relationship with government agencies across a number of issue areas. In pluralist systems, by contrast, firms are more likely to adopt transactional approaches, where engagement with government is on an issue-by-issue basis.

While the CPS literature views the firm a political actor, and one generally antagonistic toward environmental regulation, the rapidly growing stream of writing on corporate environmental management explores the mutual economic and environmental value of pursuing "green" strategies (Klassen and McLaughlin 1996; Kolk 2000; Reinhardt 2000). Various sources of gain are posited, and many advocates argue passionately that substantial "win-win" opportunities exist for the simultaneous pursuit of financial and environmental goals. For example, the application of total quality and lean production management techniques offers the potential for reducing pollution, energy usage, and waste at their source within the production process, rather than at the "end of the pipe" (Fischer and Schot 1993). Writers offer many examples where the redesign of products and production processes has reduced pollution, while simultaneously reducing fuel and material expenses and the costs of waste disposal, insurance, legal fees, and liability (Smart 1992). In addition, skillful marketing of green products can generate positive publicity and create attractive new market segments with premium prices (Coddington 1993).

These literatures on corporate political strategy and environmental management enrich our understanding of corporate practice at the firm level, but they tend to be decontextualized from the wider relations of power and have missed opportunities to engage with international environmental regimes. Here we use Gramsci's multilevel analysis to build a coherent framework that can link the macroworld of international governance structures with the microlevel of corporate practices within specific environmental issue arenas. According to Aronowitz (1981, 167), Gramsci's theory of the historical bloc can be applied to contemporary politics by "building from a micropolitics of autonomous opposition movements, whether derived from production relations or not." Such movements might include feminism, environmentalism, racial and ethnic groupings, and their motivations can extend beyond economic concerns to include identity and social legitimacy, as argued by theorists of "new social movements" (McAdam, McCarthy, and Zald 1996).

One early application of Gramscian "micropolitics" was in labor process theory, which examines the political dimensions of managerial strategies to secure labor's cooperation at the workplace. Burawoy's classic work (1979) cited Gramsci's notes on Americanism and Fordism to explain the "manufacture of consent" at the site of production:

it was relatively easy to rationalize production and labour by a skilful combination of force (destruction of working-class trade unionism on a territorial basis) and persuasion (high wages, various social benefits, extremely subtle ideological and political propaganda) and thus succeed in making the whole life of the nation revolve around production. Hegemony here is born in the factory. (Gramsci 1971, 285)

Labor process theory has extended these insights to consider technical control rooted in the production process (Braverman 1974), bureaucratic control (Edwards 1979), and the "concertive" control of total quality management and teamwork (Barker 1993). These strategies, combined with the increasing use of contingent workers and offshore sourcing, have material and ideological dimensions, rooted in production, surveillance, and disciplinary techniques (Smith and Thompson 1998).

Gramsci's conception of hegemony thus provides a basis for a more critical approach to corporate political strategy that emphasizes the interaction of material and discursive practices, structures, and strategems in sustaining corporate dominance and legitimacy in the face of environmental challenges. Corporations practice strategy to improve their market and technological positioning, sustain social legitimacy, discipline labor, and influence government policy. Shrivastava describes the "continuing political battles that proactively shape the structure of competition," and emphasizes the need to analyze "the social and material conditions within which industry production is organized, the linkages of economic production with the social and cultural elements of life, the political and regulatory context of economic production, and the influence of production and firm strategies on the industry's economic, ecological, and social environments" (Shrivastava 1986, 371–374). A key implication of these linkages is that the traditional distinction between conventional (market) and political (nonmarket) strategy is untenable (Callon 1998; Granovetter 1985); all strategy is political, in this broader sense. Moreover, the significance of coordinating the deployment of economic, political, and discursive strategies suggests a strategic concept of power, opening space for groups with fewer material resources to challenge the hegemonic position of those with structural advantages.

This broader conception of corporate political strategy provides a more nuanced understanding of the rise of corporate environmental management (CEM). While proponents claim that firms have dramatically "changed course," as alleged in the title of Schmidheiny's influential book (1992), many environmentalists have tended to view the CEM phenomenon as, at best, managerialist incrementalism, or dismiss it outright as tokenistic "greenwashing" (Rowell 1996). A Gramscian sensitivity suggests that CEM is not just a set of corporate practices, but also represents a political response to growing public and regulatory pressure over environmental problems. On the practical, material level, CEM can address some of the more flagrant environmental consequences of industrial production, while positioning companies to take advantage of new markets created by regulation or "green" consumers. On the ideological and symbolic level, CEM portrays a fundamental harmony of economic and environmental interests by constructing products as "green" and depicting firms as responsible stewards of the environment (Bansal and Roth 2000; Purser, Park, and Montuori 1995). This is more than just cynical public relations; corporate managers often come to internalize the "win-win" discourse (Levy and Rothenberg 2002). Together with more overtly political measures, such as lobbying governments and forming alliances with environmental organizations, CEM represents a series of strategies and accommodations that help to shore up corporate legitimacy and autonomy and deflect the threat of more drastic regulation. It is thus more about political and economic than environmental sustainability.

Corporate environmental management highlights several key themes in the neo-Gramscian framework, particularly the close linkages between material and ideological strategies and hence the broadly political nature of market as well as nonmarket responses. It suggests that the growth of private environmental standards and certification schemes are part of a broader trend toward private governance (Clapp, this volume; Cutler, Haufler, and Porter 1999). Similarly, the phenomenon of business-NGO partnerships should not be seen uncritically as a demonstration of harmonious interests, but rather as part of the struggle for legitimacy and influence within civil society (Ford, this volume; Murphy and Bendell 1997). The framework also illustrates the dynamic sets of moves and countermoves that sometimes enable challengers to exploit concessions in order to make further gains. For example, corporate adoption of the rhetoric of sustainability has enabled environmentalists to call attention to discrepancies between public relations and reality. Corporate concessions over the reporting of Toxic Release Inventory data in the United States enabled environmental groups to construct league tables showing the worst performers and exert further pressure to reduce emissions.

Above all, the conceptual framework developed here provides a bridge between environmental practices and strategies at the firm level, and the development of international environmental regimes of governance. In the ozone case, corporate technological strategies were both a response to emerging environmental concerns and an important driver of the particular form and timing of the Montreal Protocol (Falkner, this volume). Similarly, the international negotiations to address genetically modified organisms stemmed from corporate market strategies to develop and market seeds based on this technology, and corporate political strategies that blocked effective oversight and regulation in the United States. A full account of the evolution of these regimes of environmental governance only makes sense in the context of the political and product environmental strategies of leading firms (Andree, this volume).

The networks of actors and concomitant material and discursive structures related to specific issue arenas closely resemble the "organizational fields" discussed in institutional theory (DiMaggio and Powell 1991; Scott and Meyer 1994), particularly those renditions that attempt to integrate aspects of the "old" institutional theory, with its attention to power and alliances (Pfeffer and Salancik 1978), with the new emphasis on legitimacy and norms (Brint and Karabel 1991). According to Hoffman's (1999, 351) analysis of environmental management in the U.S. chemical industry, organizational "fields become centers of debates in which competing interests negotiate over issue interpretation." Fligstein (1996) explicitly uses the "markets as politics" metaphor as a conceptual tool for analyzing internal battles for corporate control and external competition for market domination. In Gramscian terms, field-level politics can fruitfully be viewed as a "war of position," a contested process of assembling and stabilizing a historical bloc. Similarly, the establishment of hegemony is equivalent to the process of field stabilization. Firms, governmental agencies, NGOs, and intellectuals seek to build coalitions that can establish policies, norms, and institutions that structure the field in particular ways. Business, unable to rely on economic power or governmental connections alone, needs support from a broader group of actors. Where institutional theory emphasizes pressures for convergence and stability, however, a Gramscian framework highlights disequilibrium and change (Levy and Kolk 2002; Levy and Rothenberg 2002). Contradictions, competing ideologies, and active agents ensure that the terrain of economic and political contestation is transformed continuously.

The neo-Gramscian framework holds the potential to reinvigorate the bargaining theory of MNC-host country relations. This theory, developed by international business scholars in the 1970s (Fagre and Wells 1982; Vernon 1971), steered between contentious debates of the day concerning the impact of MNCs on host countries, and instead offered a contingent approach to understanding the power relations between states and MNCs. According to the bargaining model, the distribution of benefits from foreign direct investment was contingent on relative bargaining power, which in turn was a function of the specific assets and capabilities held by each side. Over a period of time, the shifting balance of power between the MNC and the host country would tend to make the original bargain obsolete, leading to some renegotiation of terms (Moran 1985).

In our framework, actors bargain over the very structures and processes of international governance. Regime structures and processes therefore reflect the power, resources, preferences, and strategies of the various actors in these contests. The uneven outcomes of these negotiations among national states, business, and civil society, over a series of specific issue arenas, account for the fragmented and untidy form of global governance. Preferences of MNCs for regime structures depend on their perceptions of their influence relative to NGOs and other protagonists in various fora. While MNCs generally favor multilateral market enabling institutions such as the WTO, Levy and Egan (1998), in a study of the climate change negotiations, argued that U.S. energy-related businesses attempted to keep any regulation at the national level where they could exercise powerful domestic influence. U.S. industry considered itself relatively weak in the international negotiations, which involved more than 140 countries and a set of international institutions, particularly those responsible for scientific assessments, with a degree of autonomy and legitimacy that provided some insulation from the interests of particular countries or industry sectors. This case, of course, stands in stark contrast to the strong international ozone regime, which received support for competitive reasons from a key group of chemical companies who were able to exert leadership in a highly concentrated industry.

Our framework differs from the traditional investment bargaining approach in several key respects (Levy and Prakesh 2003). First, where the traditional model assumed that only MNCs and states participate in negotiations, a model of bargaining in international governance needs to take account of multiple actors, including elements of civil society. Indeed, even states may be represented by multiple authorities, such as departments of environment and industry, with conflicting interests. Second, the traditional model emphasized the economic dimension of bargaining power arising from access to unique resources, while the new model points to the importance of discursive and cultural power to frame debates in specific ways, and the significance of organizational capacity and alliances. These sources of power are not simply additive; the interplay of material, discursive, and organizational resources in a "war of position" is critical to success. Similarly, where interests were seen as objective and primarily economic, our framework suggests that interests are themselves constructed in institutional contexts and subject to political contestation. Finally, in the neo-Gramscian perspective bargaining is a complex, dynamic, and somewhat indeterminate affair. While the traditional bargaining framework did embody a dynamic element that considered shifts in power and the pressure to revise agreements, each bargain was seen as a discrete event occurring at a specific point in time, and linkages across separate bargains were rarely examined. In the framework developed here, regime development is an ongoing path-dependent process, sometimes extended over many years. Linkages also exist across regimes; for example, the institutional arrangements for providing scientific assessments to the ozone negotiations became the reference point for the climate regime. More broadly, the market-based norms of dominant regimes such as the WTO have informed the premises of regulatory regimes. As a result, while firms may not always obtain everything they want, bargaining is not pluralist competition among equals.

Conclusions

A synthesis of the macrolevel political economy perspectives with the more microlevel approaches from organization and management theory provides the basis for a conceptual framework for understanding the role of business in international environmental governance. The development of individual regimes is constitutive of the broader system of governance, yet simultaneously constrained by it. The terrain of bargaining among MNCs, states, and NGOs over each environmental regime has unique features associated with the particular set of actors and institutions involved, the structure of competition, scientific understandings and public perceptions of risks. Nevertheless, all regimes are developing in a broader system of environmental governance with a common set of norms and expectations regarding, for example, the role of market instruments and private initiative.

It is useful to highlight some distinctive contributions that a neo-Gramscian perspective brings to regime analysis. First, we might expect to observe specific strategies as actors engage in a "war of position" across the three pillars of hegemony. On the material level, companies develop product and technology strategies to secure existing and future market positions. On the discursive level, companies attempt to challenge the scientific and economic basis for regulation and use public relations to portray themselves and their products as "green," adopting the language of sustainability, stewardship, and corporate citizenship. On the organizational level, companies build issue-specific coalitions that traverse sectoral and geographic boundaries and reach into civil society. Second, and more significantly, the framework brings intellectual coherence and a more critical understanding to a number of areas of organizational theory with a bearing on environmental governance, including corporate political strategy, environmental management, bargaining theory, and institutional theory. It also places all of them in broader, more political context.

The framework is encouraging for environmental NGOs because it points to the potential for outmaneuvering rivals through the use of sophisticated analysis and strategy, good timing, and some luck. NGOs are sometimes able to compensate for their lack of resources by co-ordinating their efforts, appealing to moral principles, and exploiting tensions among states and industry sectors with divergent interests. Such strategic opportunities are likely to be more prevalent when issues are highly complex, with multiple actors, contingencies, and issue linkages, because it becomes more difficult to foresee the consequences of actions and to exercise power in conventional ways. Nevertheless, hegemony, by its nature, is resilient. As Gramsci acknowledged, dominant actors often attempt to absorb social pressures and protect their position through an accommodationist strategy of "passive revolution." The practices and discourse of corporate environmental management can be understood in these terms.

The neo-Gramscian framework suggests that stable and effective inter-national regimes require the formation of an historical bloc in both senses of the term: first, an alliance among states, leading business sectors, NGOs, and assorted professionals; second, an alignment of economic, organizational, and ideological forces that coordinate the interests of the members of the bloc. These processes are most clearly illustrated in this volume by the studies on climate change and bio-technology. The contested and contingent nature of Gramsci's notion of hegemony finds a path between state-centered accounts of traditional regime theory and overly instrumental accounts of corporate power. The process of forming an historical bloc also accounts well for the dynam-ics of issue development. Even in the absence of external shocks, a series of minor developments in the economic, discursive, and organizational realms can lead to instability and change. The rapid movement of com-panies in fossil-fuel-related industries toward a more accommodating stance after 1997 cannot simply be explained in terms of new scientific discoveries or technological changes, but flowed from a cascading sequence of events endogenous to the issue arena, in which actors' inter-ests evolved along with their strategies (Levy and Egan 2003). While the indeterminacy of complex negotiations makes it impossible to predict the precise form of an environmental regime, a detailed analysis of industry structures and competitive dynamics helps to provide insight into the reasons why specific mechanisms and structures evolve in the context of a particular environmental issue.

Notes

1. This chapter builds on an article written by the two editors that appeared in *Global Environmental Politics*, 2(4), 2002.

2. We use the term "neo-Gramscian" in acknowledgment that our conceptual framework does not rely on Gramsci's writing in any doctrinaire sense and that it also owes intellectual debts elsewhere.

References

Aronowitz, S. (1981). *The crisis in historical materialism: Class, politics, and culture in Marxist theory*. Minneapolis: University of Minnesota Press.

Bansal, P., & Roth, K. (2000). Why companies go green: A model of ecological responsiveness. *Academy of Management Journal*, 43(4), 717–736.

Barker, J. R. (1993). Tightening the iron cage: Concertive control in self-managing teams. *Administrative Science Quarterly*, 38(3), 408–438.

Baron, D. P. (1997). Integrated strategy, trade policy, and global competition. *California Management Review*, 39(2), 145–169.

Boddewyn, J. J., & Brewer, T. L. (1994). International-business political behavior: New theoretical directions. *Academy of Management Review*, 19(1), 119–143.

Boehmer-Christiansen, S. A. (1996). The international research enterprise and global environmental change. In J. Vogler & M. Imber (Eds.), *The environment and international relations* (pp. 171–175). London: Routledge.

Braverman, H. (1974). *Labor and monopoly capital: The degradation of work in the twentieth century*. New York: Monthly Review Press.

Brint, S., & Karabel, J. (1991). Institutional origins and transformations: The case of American community colleges. In W. W. Powell & P. J. DiMaggio (Eds.), *The new institutionalism in organizational analysis* (pp. 337–360). Chicago: University of Chicago Press.

Burawoy, M. (1979). *Manufacturing consent: Changes in the labor process under capitalism*. Chicago: University of Chicago Press.

Callon, M. (1998). *The laws of the markets*. Oxford: Blackwell.

Coddington, W. (1993). *Environmental marketing*. New York: McGraw-Hill.

Cox, R. F. W. (1987). *Production, power, and world order*. New York: Columbia University Press.

Cox, R. W. (Ed.) (1996). *Business and the state in international relations.* Boulder, CO: Westview Press.

Cutler, C. A., Haufler, V., & Porter, T. (Eds.) (1999). *Private authority and international affairs.* Albany, NY: SUNY Press.

DeSombre, E. R. (2000). *Domestic sources of international environmental policy: industry, environmentalists, and U.S. power.* Cambridge, MA: MIT Press.

DiMaggio, P., & Powell, W. (Eds.) (1991). *The new institutionalism in organizational analysis.* Chicago: University of Chicago Press.

Edwards, R. (1979). *Contested terrain: The transformation of the workplace in the twentieth century.* New York: Basic Books.

Fagre, N., & Wells, L. T. (1982). Bargaining power of multinationals and host governments. *Journal of International Business Studies, 13,* 9–23.

Fischer, K., & Schot, J. (Eds.) (1993). *Environmental strategies for industry.* Washington D.C.: Island Press.

Fligstein, N. (1996). Markets as politics: A political cultural approach to market institutions. *American Sociological Review, 61*(4), 656–673.

Germain, R. D., & Kenny, M. (1998). Engaging Gramsci: International theory and the new Gramscians. *Review of International Studies, 24*(1), 3–21.

Getz, K. A. (1997). Research in corporate political action: Integration and assessment. *Business and Society, 36*(1), 32–73.

Giddens, A. (1984). *The constitution of society: Outline of the theory of structuration.* Berkeley, CA: University of California Press.

Gill, S. (1995). Globalisation, market civilisation, and disciplinary neoliberalism. *Millenium: Journal of International Studies, 24*(3), 399–423.

Gladwin, T. N., & Walter, I. (1980). How multinationals can manage social and political forces. *Journal of Business Strategy,* (summer), 54–68.

Gramsci, A. (1971). *Selections from the prison notebooks* (Q. Hoare and G. Nowell-Smith, Eds. and Trans.). New York: International Publishers.

Gramsci, A. (1995). *Further selections from the prison notebooks* (D. Boothman, Trans.). Minneapolis: University of Minnesota Press.

Granovetter, M. (1985). Economic action and social structure: The problem of embeddedness. *American Journal of Sociology, 91,* 481–510.

Hall, S. (1986). Gramsci's relevance for the study of race and ethnicity. *Journal of Communication Inquiry, 10*(2), 5–27.

Helvarg, D. (1994). *The war against the greens.* San Francisco, CA: Sierra Club Books.

Hillman, A. J., & Hitt, M. A. (1999). Corporate political strategy formulation: A model of approach, participation, and strategy decisions. *Academy of Management Review, 24*(4), 825–842.

Hoffman, A. J. (1999). Institutional evolution and change: Environmentalism and the U.S. chemical industry. *Academy of Management Journal, 42*(4), 351–371.

Klassen, R. D., & McLaughlin, C. P. (1996). The impact of environmental management on firm performance. *Management Science, 42*, 1199–1213.

Kolk, A. (2000). *Economics of environmental management.* New York: Financial Times.

Lehne, R. (1993). Industry and politics: The United States in comparative perspective. Englewood Cliffs, NJ: Prentice Hall.

Leone, R. A. (1986). *Who profits: Winners, losers, and government regulation.* New York: Basic Books.

Levy, D. L., & Egan, D. (1998). Capital contests: National and transnational channels of corporate influence on the climate change negotiations. *Politics and Society, 26*(3), 337–361.

Levy, D. L., & Egan, D. (2003). A neo-Gramscian approach to corporate political strategy: Conflict and accommodation in the climate change negotiations. *Journal of Management Studies, 40*(4), 803–830.

Levy, D. L., & Kolk, A. (2002). Strategic responses to global climate change: Conflicting pressures on multinationals in the oil industry. *Business and Politics, 4*(3), 275–300.

Levy, D. L., & Prakesh, A. (2003). Bargains old and new: Multinationals in international governance. *Business and Politics, 5*(2), 131–151.

Levy, D. L., & Rothenberg, S. (2002). Heterogeneity and change in environmental strategy: Technological and political responses to climate change in the automobile industry. In A. Hoffman & M. Ventresca (Eds.), *Organizations, policy and the natural environment: Institutional and strategic perspectives* (pp. 173–193). Stanford, CA: Stanford University Press.

Lipschutz, R. D. (1992). Reconstructing world politics: The emergence of global civil society. *Millenium, 21*(3), 389–420.

Mahon, J. (1983). Corporate political strategies: An empirical study of chemical firm responses to superfund legislation. In L. Preston (Ed.), *Research in corporate social performance and policy,* Vol. 5 (pp. 143–182). Greenwich, CT: JAI Press.

McAdam, D., McCarthy, J., & Zald, M. (1996). *Comparative perspectives on social movements.* New York: Cambridge University Press.

Moran, T. H. (Ed.) (1985). *Multinational corporations: The political economy of foreign direct investment.* Lexington, MA: Lexington Books.

Murphy, C. N. (1998). Understanding IR: Understanding Gramsci. *Review of International Studies, 24*(3), 417–425.

Murphy, D. F., & Bendell, J. (1997). *In the company of partners: Business, environmental groups, and sustainable development.* Bristol: Policy Press.

Oster, S. M. (1982). The strategic use of regulatory investment by industry subgroups. *Economic Inquiry, 20*, 604–618.

Pfeffer, J., & Salancik, G. R. (1978). *The external control of organizations: A resource dependence perspective.* New York: Harper and Row.

Purser, R. E., Park, C., & Montuori, A. (1995). Limits to anthropocentrism: Towards an ecocentric organization paradigm? *Academy of Management Review, 20*(4), 1053–1089.

Reinhardt, F. L. (2000). *Down to earth: Applying business principles to environmental management.* Boston: Harvard Business School Press.

Robinson, W. (1996). *Promoting polyarchy: Globalization, U.S. intervention, and hegemony.* New York: Cambridge University Press.

Rowell, A. (1996). *Green backlash: Global subversion of the environmental movement.* London: Routledge.

Rupert, M. (1998). (Re-) Engaging Gramsci: A response to Germain and Kenny. *Review of International Studies, 24*(3), 427–434.

Schmidheiny, S. (1992). The business logic of sustainable development. *The Columbia Journal of World Business, 27*, 18–25.

Schreurs, M. A. (Ed.) (1997). *The internationalization of environmental protection.* Cambridge: Cambridge University Press.

Schuler, D. (1996). Corporate political strategy and foreign competition: The case of the steel industry. *Academy of Management Journal, 39*(3), 720–737.

Scott, W. R., & Meyer, J. W. (Eds.) (1994). *Institutional environments and organizations.* Thousand Oaks, CA: Sage.

Shaffer, B. (1992). Regulation, competition, and strategy: The case of automobile fuel economy standards 1974–1991. In J. Post (Ed.), *Research in corporate social performance and policy* (Vol. 13, pp. 191–218). Greenwich, CT: JAI Press.

Shrivastava, P. (1986). Is strategic management ideological? *Journal of Management, 12*, 363–377.

Skidmore-Hess, D. (1996). Business conflict and theories of the state. In R. W. Cox (Ed.), *Business and the state in international relations* (pp. 199–216). Boulder, CO: Westview Press.

Sklair, L. (1997). Social movements for global capitalism: The transnational capitalist class in action. *Review of International Political Economy, 4*(3), 514–538.

Smart, B. (1992). *Beyond compliance.* Washington, DC: World Resources Institute.

Smith, C., & Thompson, P. (1998). Re-evaluating the labour process debate. *Economic and Industrial Democracy, 19*(4), 551–577.

van der Pijl, K. (1998). *Transnational classes and international relations.* London: Routledge.

Vernon, R. (1971). *Sovereignty at bay: The multinational spread of U.S. enterprises*. New York: Basic Books.

Wapner, P. (1995). Politics beyond the state: Environmental activism and world civic politics. *World Politics, 47,* 311–340.

Weidenbaum, M. L. (1981). *Business, government, and the public*. Englewood Cliffs, NJ: Prentice Hall.

Young, O. R. (1994). *International governance: Protecting the environment in a stateless society*. Ithaca, NY: Cornell University Press.

II

Business Strategies in International Environmental Governance

4

Business and the Evolution of the Climate Regime: The Dynamics of Corporate Strategies

David L. Levy

The prospect of mandatory curbs on production and use of fossil fuels to halt potentially catastrophic changes in the Earth's biosphere poses a serious economic threat to corporate actors across a wide range of industries. Carbon dioxide is the single most important greenhouse gas (GHG) and is released through the combustion of fossil fuels. Newell and Paterson (1998) conclude that "when the centrality of fossil fuels in producing global warming is combined with the centrality of fossil energy in industrial economies, it becomes clear that the fundamental interests of major sectors of those economies are threatened by proposals to limit greenhouse gas emissions." The fossil fuel industry's initial response to this threat varied considerably across regions and industries. U.S. companies in the coal, oil, automobile, utility, and chemicals industries formed industry associations, lobbied politicians, challenged the science of climate change, and pointed to the high costs of reducing emissions. European industry, by contrast, was far less aggressive in responding to the issue, and displayed a greater readiness to invest in technologies that might reduce greenhouse gas emissions. Responses also varied across sectors, with the oil industry appearing less willing to countenance mandatory emission controls than the automobile industry in both the United States and Europe.

By 2000, key firms in a number of industry sectors and on both sides of the Atlantic appeared to be converging toward a more accommodative position that acknowledged the role of GHGs in climate change and the need for some action by governments and companies, despite continuing uncertainty. In the oil and automobile industries, companies were beginning to invest substantial amounts in low-emission technologies, and were engaging a variety of voluntary schemes to inventory, curtail, and trade carbon emissions. No obvious dramatic scientific, technological, or regulatory developments can account for these changes; investments in low-emission products and processes remained highly risky, and the prospects for ratification in the U.S. Senate of the 1997 Kyoto Protocol were slim.

This chapter focuses on corporate responses to climate change in the oil and automobile industries, exploring the differences between U.S.-based and European multinational corporations (MNCs) and changes in their positions over time. It is based on research and interviews conducted with four oil and four automobile companies during 1998–2000, half of which were based in the US and half in Europe. The chapter makes two major arguments. First, it attempts to locate the response strategies of companies not just in terms of their market and competitive positioning, but also in reference to their histories and institutional environments. Corporate strategies are derived from perceptions of economic interest that are mediated by the different cultural, political, and competitive landscapes in the United States and Europe. Expectations concerning markets, technologies, regulatory responses, consumer behavior, and competitor reactions varied among the companies according to their individual histories, headquarters location, and membership in particular industry organizations, partly explaining the divergent corporate responses. The recent shift by many companies toward a more accommodating stance, it is argued, is related to convergent pressures at the level of the global auto and oil industries, as well as the influence of the climate issue arena itself. Interactions among actors during the extended negotiations have tended to narrow the range of corporate perspectives on climate change. This shift in perspectives has combined with changing competitive dynamics, the evolution of new organizations supportive of a proactive industry role, and the diffusion of "win-win" discourse articulating the consonance of environmental and business interests.

The second major claim of this chapter is that the emergence of a climate regime can best be characterized as a contested process of assembling an historical bloc, in both senses in which Gramsci used the term. At one level, actors engage in negotiation, alliance formation, and compromise, in an effort to build a hegemonic coalition of firms, governmental agencies, NGOs, and intellectuals with the capacity to establish policies, norms, and institutions that structure the regime in particular ways. The Gramscian approach reflects the negotiated nature of international environmental agreements; even the most powerful states are generally unable to impose a particular agreement on the international community, though they may be able to block or delay for some time. Similarly, business is unable to determine state policies or write the rules for environmental regimes in a direct instrumentalist manner. This is not, however, a pluralist story of interest group bargaining among equals; the broader dominance of neoliberal institutions provides business with significant leverage. The emerging climate regime illustrates both the significant role of business and the contested and contingent nature of hegemony.

On a second level, the formation of an historical bloc refers to a configuration of economic, discursive, and organizational forces that is capable of providing some cohesion and stability to the regime by aligning and coordinating actors' perceived interests. The notion of hegemony, in this sense, is similar to the concept of field stabilization in institutional theory. Participation in industry associations or national organizations informs the ways in which firms perceive their economic interests. Companies as well as countries will consent to a regime if they see that policy measures present only minor economic threats or even some opportunities. Perceptions of economic threat and opportunity depend on initial competitive locations of firms and their capacity to compete in new markets for low-emission technologies. The specific mechanisms of the regime, such as emission trading and the Clean Development Fund, might provide economic incentives that are attractive to firms with appropriate capabilities. The diffusion of the "win-win" discourse of eco-modernism also encourages firms to think that they can benefit economically from environmental investments.

For a stable regime to emerge, major actors also have to share some common frames regarding scientific understanding of the issue and policy approaches to mitigation and adaptation. It is by no means guaranteed that economic and discursive forces will provide sufficient coordination of interests to produce a hegemonic coalition capable of stabilizing a regime. The difficulty in accommodating the fossil fuel industry and energy-dependent countries such as the United States and Australia accounts for the fragility and flexibility of the Kyoto Protocol; reduction targets are very modest, and can mostly be achieved without any real reduction of domestic industrial emissions. A corollary implication is that even if a regime is established, it does not necessarily address environmental problems in any adequate way.

Bargaining over regime structures and processes engages actors in a complex set of strategic maneuvers in the economic, technological, and political spheres. Gramsci's concept of "war of position" is useful for describing a struggle coordinated across multiple bases of power to defend or advance one's position in the face of an issue such as climate change. One implication is that the traditional distinction between political and market strategies is unsustainable; any threat to an industry's markets, whether from regulation, environmental NGOs, or technological innovation, is simultaneously an economic threat and a challenge to hegemonic stability. Similarly, corporate response strategies to such threats, including research and development, mergers, or lobbying, are both economic and political in nature. Actors' interests are not given and fixed in this process; rather, they renegotiate conceptions of their own interests as a result of shifting institutional influences, technological and market developments, and strategic interactions with other actors. A regime emerges if and when this bargaining process generates economic and discursive structures that align conceptions of interest of major actors and that are relatively stable. Understanding regime formation therefore requires detailed examination of actors' strategies and their shifting perceptions of interests.

Climate Change as a Threat to Hegemony

The possibility that human emissions of greenhouse gases are changing the world's climate system constitutes a global environmental issue with massive market transforming potential. Controls on emissions of carbon dioxide (CO_2), released from the combustion of fossil fuels and the main contributor to global warming, would threaten oil and coal companies as well as industries dependent on these fuels, particularly transportation and electric utilities. In addition, higher energy prices would raise input costs for a range of energy intense industries, including aluminum, chemicals, cement, paper, cement, and steel (Mansley 1995). There is little assurance that incumbent companies would be the winners in future markets for low emission products; unlike the ozone depletion case, where DuPont successfully led the innovation and marketing of CFC-substitutes, technologies to address greenhouse gases represent radical technological innovations far from the core areas of expertise of incumbent firms (Anderson and Tushman 1990; Christensen 1997; Henderson and Clark 1990). Investments in R&D for low-GHG products and processes appear highly risky because of the uncertainty regarding climate science, regulatory responses, and the potential market for low emission technologies.

The threat from climate change extends beyond the purely economic realm to the ideological foundations of corporate legitimacy and autonomy (Levy and Egan 1998). A former vice president of Government Relations of a U.S.-based car company commented that "there are people who have cast the automobile as a villain. It is a puritanical view, that we are having too much fun, that we have too much mobility and freedom, that suburban sprawl is bad. They think we should all live in beehives. So when scientists say that CO_2 is a greenhouse gas, they jump on board".[1] Automobile industry managers expressed fear that the climate issue touched emotional chords that could be exploited by activist environmental groups. On the organizational level, the climate issue has strengthened the position of environmental groups, regulatory agencies, as well as nascent companies pursuing low-emission technologies. Such coalitions could push government agencies to tighten regulations such as U.S. CAFE (Corporate Average Fuel Economy) standards for automobiles. Moreover, the issue was already spurring the development of international institutions that monitor and address global environmental problems. The growth of international environmental assessments and negotiations has also expanded the organizational capacity and legitimacy of NGOs (Wapner 1995) and communities of scientific experts (Haas 1993). Corporate responses to climate change are therefore geared not just to economic survival in a carbon-constrained world, but also to sustaining the moral authority and

leadership essential to hegemony. The World Economic Forum, an influential annual gathering of several thousand leading business managers,
policymakers, and academics, voted climate change as the most important issue facing business at the 2000 meeting in Davos, and, according
to a press release, "Not only did the audience choose climate change
as the world's most pressing problem, they also voted it as the issue
where business could most effectively adopt a leadership role" (World
Economic Forum 2000).

Despite the common threat to companies in fossil fuel-related industries, there has been a striking variation in the responses of companies
across sectors and countries. In the oil industry, for example, U.S.-based
companies such as Exxon and Chevron have expended considerable
energies in aggressively challenging climate science, pointing to the
potentially high economic costs of greenhouse gas controls, and lobbying against mandatory international emission controls. By contrast, BP
and Shell, the two largest European companies, have proclaimed their
acceptance of the scientific basis for precautionary action and have
announced substantial investment plans for renewable energy. These
divergent strategies defy simple explanation. The more obvious economic
and technological characteristics of the companies, such as the carbon
intensity of their production and reserves, do not explain the differences
(Rowlands 2000). Indeed most of these companies are large, integrated
multinationals with similar profiles and strategic capabilities, and pos-
sessing production and distribution operations spread throughout North
America, Europe, and the Middle East.

Institutional theory posits that corporate perspectives on climate
change are likely to be mediated by the institutional environment at the
national and industry levels (Smircich and Stubbart 1985). Oliver (1997)
argued that uncertainty increases the influence of the institutional environment and reduces the impact of economic and competitive factors.
Given the high level of uncertainty surrounding the climate issue, firms
cannot easily make a rational, objective calculus of their economic interests and appropriate strategic responses, and might therefore be more
subject to institutional pressures. The sharp distinction between institutional and economic explanations breaks down, however, under closer
examination. A more useful theoretical approach avoids this dualism

and recognizes that economic calculations of interests always embody assumptions which may be more or less certain and are constructed in broader social contexts (Callon 1998).

The home country of multinational corporations (MNCs) is one important aspect of the institutional environment that can play a role in shaping corporate strategy (Sethi and Elango 1999). These institutional influences are likely to preserve the legacy of the country of origin, even in highly internationalized companies, because most MNCs still concentrate their senior management responsible for strategy development in the country of origin. Much has been made of the different cultural values and institutional norms in the United States and Europe. Kempton and Craig (1993) found that Europeans expressed more concrete concerns about environmental impacts on future generations and viewed their responsibility for sustainability as part of their national identity and heritage. People in the United States demonstrated more concern about economic costs of regulation and were optimistic about technical solutions. Empirical studies of corporate environmental management practices, however, do not support the notion that cultural differences translate into significantly different environmental practices (KPMG Environmental Consulting 1999; United Nations Transnational Corporations and Management Division 1993).

In the political arena, the American system of business-government relations is often characterized as adversarial compared to the more corporatist arrangements in Europe, where key stakeholders enter into a process of collaborative bargaining with governments (Vogel 1978). Several U.S. managers acknowledged that adopting an adversarial stance concerning climate change was almost expected and would hardly cost them much credibility with regulators; one Exxon manager stated "they cannot ignore us anyway; we are the big elephant at the table." By contrast, European managers were much more concerned with securing a voice and a "seat at the table" in the negotiations. In practice, however, business influence over policy in Europe is also very strong. The lack of public participation and transparency in the process of formulating policy means that business interests are given significant scope to define the policy agenda (Mazey and Richardson 1993; McLaughlin, Jordan, and Maloney 1993). While environmental groups may exercise influence in setting the agenda, at key decision points it is the large MNCs and industry associations that have preferential access to members of the Commission and governments in member states (Coen 1999). For example, the successful lobbying campaign against the proposed EU carbon tax was described by *The Economist* (1992) as the "most powerful offensive against an EC proposal ever mounted by Europe's industrialists." Not only is business influence in the EU pervasive, it is also increasingly influenced by the European subsidiaries of U.S.-based companies. According to Coen (1999: 27), the American firms have brought to Europe a propensity to establish alliances and issue-specific policy networks as a result of which "the government-business relationship in Brussels has developed many similarities to the type of lobbying observed in Washington, D.C."

The impact of MNCs' countries of origin on corporate perspectives and strategies is likely to diminish over time as industries become more international in scope. MNCs are increasingly delinked from their home countries and more integrated into global industries. A strong case can be made that the international oil and automobile industries constitute coherent institutional fields, with cognitive, normative, and regulatory pressures inducing some measure of convergence (Scott and Meyer 1994). For example, oil is a commodity with a uniform international price, and the major companies adopt global rather than multidomestic strategies, at least in their production and refining operations. Perhaps more than any other industry, oil companies approach strategy in an internationally coordinated manner, while utilizing global sourcing, integration, and rationalization to achieve economies of scale and low costs. Given the keen awareness of interdependence, companies are likely to copy each others' moves to prevent rivals gaining undue advantage (Chen and MacMillan 1992; Chen and Miller 1994; Knickerbocker 1973). Industry interdependence also takes a collaborative form, within industry associations and in a number of alliances and joint ventures. Executives read the same trade journals and the same studies of industry trends. The automobile industry is not far behind the oil industry in terms of globalization.

The emergence of climate change as a "global issues arena" itself constitutes an institutional context that provides some convergent pressure for companies engaged with the issue. It represents a concrete instance of the "international determinants of [MNCs]' sphere of operations," which Hirst and Thompson (1996) contend would be needed to generate convergence. MNCs have little choice but to develop unified company-wide positions toward such issues, even when some subsidiaries dissent from the corporate stance. Indeed, most of the large MNCs in the automobile and oil sectors have formed internal cross-functional "climate teams" for precisely this purpose. The network of actors involved in a global issues arena interact frequently and develop their own organizational and institutional frameworks. In the climate case, the senior managers responsible for climate-related strategy know each other well and meet regularly at the international negotiations and at other conferences and industry-level activities.

Corporate strategic responses to climate change are thus mediated by divergent pressures at the level of the individual firm and national cultural and regulatory systems, and by convergent forces at the level of the global industry and the climate change issue itself. The following analysis of the oil and automobile industries in the United States and Europe illustrates these pressures, and suggests that initial responses were primarily framed by national and individual firm pressures, accounting for divergent strategies, while later trans-Atlantic convergence can be explained in terms of the increasing influence of the global industry and the climate issue itself.

U.S. Industry Response to Climate Change

The initial response by U.S. industry was rapid and aggressive, and displayed a degree of coordination across the organizational, discursive, and economic bases of hegemony. The most immediate business response to the perceived threat of greenhouse gas was in the organizational domain. The Global Climate Coalition (GCC) was formed in 1990, representing about forty companies and industry associations who were major producers and users of fossil fuels. Although the GCC was constituted as a U.S.-based organization and was focused on domestic lobbying, a number of U.S. subsidiaries of European multinationals also joined, and the GCC quickly rose to be the most prominent voice of industry, both in the United States and internationally. During the 1990s, a number of organizational initiatives attempted to improve communication and coordination among multinational corporations (MNCs) from different countries, not always with great success. These efforts include the Business Council for Sustainable Development, which was particularly active at the UN Rio conference in 1992 (Finger 1994) and the International Chamber of Commerce (ICC). The ICC, whose membership is primarily drawn from OECD countries, has a very active working party on climate change, which met in London in January 1996 to plan strategy for the Second Conference of the Parties (COP-2) session in Geneva in July 1996.

On the discursive level, a key strategy of the GCC in its opposition to mandatory emission controls has been to challenge the science of climate change, pointing to the lack of consensus among scientists and highlighting the uncertainties. Industry often invokes the authority of scientific discourse to deflect regulation, demanding a high burden of proof of potential harm, or "sound science," before any action is taken (Jasanoff 1987). The GCC used this approach explicitly in its mission statement: "A bedrock principle for addressing global climate change issues is that science—not emotional or political reactions—must serve as the foundation for global climate policy decisions." The GCC actively promoted the views of climate skeptics in its literature, press releases, and congressional testimony, and sponsored a number of reports (AccuWeather 1994; Davis 1996). Fossil fuel interests also funded a number of "astroturf" groups, such as the "Information Council for the Environment (ICE)," founded in 1991, whose purpose was to "reposition global warming as theory, not fact" (Gelbspan 1997; Ozone Action 1996).

Industry associations have also stressed the potentially high economic cost of emission controls. The GCC commissioned a series of economic studies (Montgomery 1995; WEFA Group and H. Zinder and Associates 1996) and in a 1996 press release warned that measures to curb emissions by 20 percent "could reduce the U.S. gross domestic product by 4% and cost Americans up to 1.1 million jobs annually." In the run-up to the Kyoto conference in December 1997, industry's message shifted from the high cost of control measures to the unfair economic burden and limited environmental effectiveness of an agreement that would exclude developing countries from emission controls. The GCC channeled $13 million through the Global Climate Information Project for an advertising campaign with the slogan "It's not global and it won't work." By emphasizing the potential impact of emission controls on fuel prices and employment, the Global Climate Information Project (GCIP) secured the endorsement of a number of associations representing a spectrum of civil society, including unions, black businesses, farmers, and retired people. A similar theme was expressed by the Chairman of Chrysler in a Washington Post editorial: "we're moving toward a solution involving a massive transfer of American wealth that won't do a thing to keep the polar ice caps from melting, but would severely undermine this country's international competitiveness" (Eaton 1997).

These organizational and discursive strategies rest on material and economic foundations. The GCC is funded by some of the world's largest MNCs, enabling it to send large delegations to international negotiations, mount expensive advertising campaigns, commission reports, and donate substantial sums to Political Action Committees (PAC) to finance U.S. Congressional election campaigns (Hamilton 1998). In the United States, fossil fuel industries donated $130,000 to President Clinton's campaign in the 1995–1996 cycle and made PAC contributions to members of the Senate Energy and Natural Resource Committee totaling approximately $200,000 (Ozone Action 1996). Greenpeace International estimates that between 1991 and 1996 the oil and gas sector donated $53.4 million to U.S. election candidates and their political parties (Greenpeace International 1997).

Responses in the Oil Industry

Within the global oil industry, Exxon has taken the firmest stand against greenhouse gas controls. In addition to citing scientific uncertainties and the exclusion of developing countries from emission reduction commitments, the company warns of the dire economic consequences of Kyoto commitments (Exxon 1999). Exxon advertises its own efforts to promote internal energy efficiency, cogeneration, and funding for fuel cell research and carbon sequestration, technologies that complement fossil fuels rather than replace them. Managers interviewed expressed the view that the company's profitability, the envy of its competitors, was due to its focus on core businesses and lean cost model. They had learned from the experiment with diversification that businesses such as office products, with rapid product cycles and very different technologies, required competencies that Exxon lacked. Lee Raymond, Exxon's CEO, viewed renewables as "a waste of money . . . oil and gas will continue to be the dominant energy for the next 25 years" (The Economist 2001).

BP is widely considered to be the most responsive company on the climate issue. John Browne's landmark speech in May 1997 was the first acknowledgement in the oil industry of a scientific case for precautionary action, and BP was the first company to leave the Global Climate Coalition (GCC), the major industry association opposing emission controls. In 1997 BP established a partnership with Environmental Defense to develop an internal carbon-trading scheme, and joined the Pew Center for Global Climate Change, which advocates for early action on the issue. In 1998 the company committed to reduce internal emissions by 10 percent by 2010 relative to a 1990 baseline; indeed, BP announced in January 2003 that it had already met this goal. BP's acquisition of Amoco greatly increased its investment in solar energy, making BP-Solarex the largest photovoltaics (PV) company in world, with revenues expected to climb to $1 billion within ten years. BP sought to redefine itself strategically as an energy company and believed that competitive advantage could be secured through a positioning that is "distinctive in the eyes of governments, consumers, and regulators" (Reinhardt 2000). This new profile became explicit with the launch of a new "green starburst" logo and the slogan "BP—Beyond Petroleum" in July 2000. CEO Browne has stated that renewables could account for 5 percent of revenues by 2020, and 50 percent by 2060 (Corzine 2000).

Shell has broadly followed BP's strategy, though with a lower public profile and a broader commitment to sustainability. Shell accepted the scientific need for action on greenhouse gas emissions and has established internal emission reduction targets. Shell International Renewables was established in 1998, consolidating existing businesses but with a new commitment to invest $500 million over five years in renewables, with the expectation that the division would be profitable by that time. Shell's investments are primarily in PV, but it also contemplates a greater role for geothermal and wind, in which it sees the potential for applying existing technological competencies in geology and offshore platforms. More recently, it has invested in power generation and distribution. Shell is also repositioning itself as an energy company and its unique long-term planning scenarios envisage that renewables will account for 30–40 percent of global energy by 2060.

Texaco is a U.S.-based oil company that began to shift position in 1999 and whose stance is now closer to BP and Shell than to Exxon. Texaco's managers argued that the climate issue is driven by public and governmental pressure rather than science, and that the momentum toward mandatory controls is unstoppable. While Texaco has no renewable energy business, CEO Peter Bijur has pushed the company toward being a broad energy company. In 2000 the company acquired a holding in Energy Conversion Devices (ECD), which is established in advanced battery technologies and has strengths in PV. Texaco believes that its proprietary gasification technologies could generate hydrogen for fuel cells.

European managers expressed far more concern for their legitimacy and image than did managers of U.S. oil firms. A BP manager stated that "as a company trying to act with corporate social responsibility, is it sensible to turn a blind eye to this issue? Our response was no." Similarly, a Shell executive said that "here in Europe it can be hard to go to church and show your face, especially with the Nigerian situation all over the newspapers. There is a real concern for legitimacy and what the community thinks. There is a fight for the hearts and minds of the public; this is a long-term force affecting our business." Following the Brent Spar incident, consumer boycotts were organized in European countries and Shell's market share dropped noticeably in Germany. One of Shell's long term planning scenarios, termed People Power, discussed the risk of significant public pressure on the environment. Exxon, by contrast, saw little value in improving its image: "If we appear more green, it might get us a better seat at the policy table, but the real question is whether it would improve our access to resources and markets. BP and Shell actually attract counter-pressure for talking green but not doing enough. There is a Norwegian saying that 'the spouting whale gets harpooned.' Greenpeace has demanded than they pull out of fossil fuels altogether."

Clearly, Exxon saw less economic value in displaying environmental responsibility than did the European oil companies.

Some of the most significant differences among the companies were individual rather than related to their country of origin. BP, Shell, and Texaco believed that the acquisition of new competencies had to be done slowly and that significant first-mover advantages might accrue. Exxon, by contrast, emphasized the risks of venturing into unknown and risky territory; over time, technological and market options would become clearer and resources could be allocated accordingly. Shell managers expressed their company's historical preference for organic, internal growth, and thought that Exxon was more likely to use an acquisition strategy to secure necessary competencies. In addition, Exxon's financial performance gave it little reason to reconsider its strategy; other companies, particularly Texaco, were forced to reevaluate their positions during periods of low profitability caused by low oil prices.

Trans-Atlantic differences were also evident in the auto industry. The technological strategies of U.S.-based companies had traditionally been geared toward addressing local air quality by reducing smog precursors such as sulfur oxides (SO_x), nitrous oxides (NO_x) and hydrocarbons. This could be achieved largely through end-of-the-pipe approaches such as improved catalytic converters rather than higher fuel efficiency. The companies were also investing in a range of alternative fuel programs for fleet vehicles, such as compressed natural gas and ethanol, which only reduce greenhouse gas emissions substantially if obtained from renewable sources such as biomass. Indeed, during the 1990s, any technological improvements on the fuel efficiency front were more than offset by increasing weight of vehicles and larger engine size (Stoffer 1997).

The main problem confronting the U.S. auto industry is that it has traditionally derived most of its profits from larger vehicles, light trucks, and more recently, sport utility vehicles (Bradsher 1999). With the majority of sales still flowing from the U.S. market, American car manufacturers have developed capabilities to fit the national environment. Low fuel prices provide little incentive for American consumers to care about fuel consumption, and the U.S. companies have rarely enjoyed much success with smaller, fuel efficient cars. The main concern, then, for GM and Ford, is less about their expertise in low-emission technologies such as advanced diesel or hybrid-electric drive trains, and more about their ability to integrate these technologies into smaller vehicles and successfully mass produce and market them. Moreover, holding the U.S. car companies back appears to be a pervasive internalisation of pessimistic perceptions and expectations concerning the market for low-emission technologies (Levy and Rothenberg 2002). For example, they appear convinced that high-efficiency diesel technology would not be accepted by Americans because consumers remember these cars as noisy, dirty, vibrating, and slow. European companies, on the other hand, are trusting that a new generation of cleaner, smoother, quieter, and more powerful direct-injection diesel cars will go a long way toward meeting emission reduction targets negotiated with the EU. Similarly, U.S. companies are locked on the idea that low-emission vehicles will not find consumer acceptance if there are any trade-offs with other vehicle attributes, such as size, range, comfort, or safety, which places all the burden of emission reduction on power train technology development. In Europe, by contrast, car companies anticipate that consumer tastes and behavior patterns may evolve to accommodate smaller, lighter, low emission vehicles, allowing greater flexibility in achieving emission reductions in the medium term.

The European automobile industry found itself in a weak political position, in which politicians were looking to the industry for substantial, early emission reductions. One reason for this was the relatively late response by industry to climate change and the lack of a broad coalition comparable to the GCC in the United States. Germany, with a well organized green political party, had unilaterally committed to significant GHG reductions during the negotiations in Berlin in 1994 and had pushed the German auto industry association into a "voluntary" agreement to reduce CO_2 emissions from new cars by 25 percent. Concerned about the competitiveness of its national automobile companies, Germany then pushed the European Union (EU) to adopt similar measures. The EU was sensitive to accusations from the Americans that it talked a tough position but lacked any will to implement. The European Commission introduced a proposal to reduce average CO_2 emissions from new cars from 186 grams/km to 120 g/km by 2005 (equivalent to about 45 mpg). European automobile companies avoided direct challenges to the scientific need for emission controls, with various managers calling any such effort "futile" and "inappropriate." After three years of negotiations, during which major rifts between the European oil and automobile industries became apparent, the European Automobile Industry Association (ACEA) accepted a voluntary agreement in 1998 to reduce emissions to 140 g/km by 2008 (Bradsher 1998). The agreement included Ford and GM's European subsidiaries, but not Japanese manufacturers. These negotiations highlighted the difference between the technocratic policy process in the United States and the lack of technical capacity of the EU institutions.

European companies have responded to the climate issue by introducing very small, lightweight cars such as Daimler-Chrysler's SMART car, and investing substantial amounts in a range of technologies from diesel to fuel cells; by 2001, over 30 percent of new passenger vehicles were diesels, attaining fuel economy rivaling hybrid vehicles. Daimler has aggressively pursued fuel cell technology, investing $320 million in the Canadian company Ballard in 1997, and has announced plans for a limited commercial launch by 2004, by which time it would have spent an estimated $1.4 billion on fuel cell vehicles (Hoffman 1999). European R&D efforts were thus more balanced and incremental than those in the United States, preparing them for a carbon-constrained world in the short to medium term.

These corporate responses were conditioned by the existing regulatory context. In the United States, the primary concern for many years had been local air quality. U.S. industry was already subject to CAFE standards under the Clean Air Act and its amendments, and the California Air Review Board (CARB) was mandating very low SO_x, NO_x, and hydrocarbon emissions. This pressure had led to precise electronic combustion control and catalytic converter technology. Initially, U.S.-based companies understood climate change as a continuation of this pressure, thus not requiring a major strategic change in direction. Helen Petrauskas, a Ford vice president, commented that "there was already huge pressure for reduction of smog precursors. So climate did not require a step function change in strategy; it was more of an organic evolution." Over a period of time, however, companies came to appreciate that many technological approaches involved trade-offs. Electric vehicles, for example, can account for substantial indirect emissions depending on the fuel mix and efficiency of electricity generation. Similarly, the introduction of catalytic converters in the early 1980s caused a noticeable decrease in fuel efficiency. It was not easy, however, for American companies to shift their technology strategies toward carbon reduction, because the fragmented regulatory system was still ratcheting up controls on non-GHG emissions while paying no attention to CO_2. In Europe, by contrast, transportation policy had historically led to high fuel prices, and European innovation efforts were therefore directed toward lighter weight, smaller vehicles with high fuel efficiency. Companies thus were more prepared for the challenge of climate change.

As in the oil industry, historical experience shaped company perspectives on the benefits of being a first mover in emerging low-emission technologies. GM, for example, had invested more than $1 billion in its electric vehicle, but sold fewer than 1,000 cars (Lippert 1997; Shnayerson 1996). Company managers generally interpreted the experience as a commercial mistake. Ford had invested an estimated $500 million in sodium-sulfur batteries, only to abandon the project because of safety concerns. With this shared experience, American companies focused on the risks of being a first mover.

Instability and Change

The coordinated strategy by the U.S. fossil fuel industry succeeded in securing powerful political allies. The United States opposed mandatory international GHG emission controls until 1996, and although the United States was a party to the 1997 Kyoto protocol, prospects for ratification by the United States were very slim. By mid-1999, however, a number of writers had noted an apparent sea change in industry's stance on climate, as companies began to accept the scientific basis for emission controls and to invest significant sums in low carbon technologies (Nauss 1999; *Newswire* 1999). This change is somewhat surprising, given that no major scientific breakthroughs had occurred, the United States was moving away from any international agreement, and no implementation mechanisms for the Kyoto Protocol had been agreed.

A Gramscian analysis suggests that the shift in the American industry's position was attributable to a series of strategic miscalculations, interactions with events in Europe, and shifts in the discursive, organizational, and economic spheres. Together, these developments opened up some of the tensions and contradictions in the historical bloc opposing mandatory emissions reductions and provided opportunities for challenges. Industry was forced into a series of accommodations that are laying the foundation for a more informal and decentralized climate regime acceptable to a broad range of actors, based on limited targets for emission reductions, flexible market-based implementation mechanisms, and minimal regulatory intrusion upon corporate autonomy.

The U.S. fossil fuel industry's opposition to mandatory controls was not well entrenched in civil society. The efforts at "astroturf organizing" only highlight the lack of real grassroots support in civil society for industry's stance. While the industry enjoyed strong support in the U.S. Congress, the Clinton administration was almost paralyzed by internal dissent over the economic impact of emission reductions, with the Department of Energy generally supportive of industry's position, the Environmental Protection Agency opposed, and Commerce split. A breach in industry ranks also became evident in the pre-Kyoto period. On June 8, 1997, the Business Roundtable sponsored full-page advertisements in the U.S. press signed by 130 CEOs, arguing against mandatory emissions limitations at the forthcoming Kyoto conference. Eighty Business Roundtable members did not endorse the advertisements, however. Monsanto had led an unsuccessful effort to draft an alternative text, which acknowledged that sufficient scientific evidence had accumulated to warrant concern and industry's engagement in developing precautionary measures. This dissenting view was brought to President Clinton's attention at the June 1997 meeting of the President's Council of Advisers for Science and Technology (PCAST), leading to the establishment of an interagency task force.

The fossil fuel industry's strategies were not always effective and sometimes had unanticipated consequences. One example was the attempt to undermine the credibility and legitimacy of the Intergovernmental Panel on Climate Change, the group of scientists conducting periodic reviews of the causes, impacts, and possibilities of mitigation of climate change, an emerging formation within global civil society. In May 1996, three industry associations publicly accused two lead IPCC authors of secretly altering the IPCC Second Assessment Report to reduce the expression of uncertainties. Despite the effectiveness of this campaign in U.S. domestic politics, the challenge to IPCC's credibility fell short; industry challenges to science were viewed much more skeptically in the international negotiations. Moreover, the U.S. delegation was forced to distance itself publicly from the fossil fuel lobby (Wirth 1996).

After this affair, industry became more circumspect about challenging the science of climate change. In the run-up to Kyoto in December 1997, the GCC shifted its strategy from highlighting scientific uncertainties toward the high costs of mitigation and the lack of developing country commitments. This decision was based, in part, upon market research that indicated the public was not engaged with the scientific debates and did not find industry a credible source. Moreover, challenging the science was producing a backlash. Environmental groups in Europe and the United States issued a number of reports that noted industry support for some climate skeptics, and attempted to frame the issue as big business using its money and power to distort the scientific debate (Corporate Europe Observatory 1997; Friends of the Earth International 1997; Gelbspan 1997; Hamilton 1998). A Ford executive acknowledged that "appearing negative hurts. We lost the first round of battles. We are now trying to be more positive with the science, while still pointing to the high cost of precipitate action before scientific uncertainties are resolved. Our actions will be less strident in the future."

The American Petroleum Institute (API) dissented from this decision, preparing a new "Global Climate Science Communications Plan" to enroll a group of climate skeptics who were not previously identified with the fossil fuel lobby. In internal documents leaked to the National Environmental Trust and the *New York Times* (National Environmental Trust 1998), the API expressed concern that the U.S. media conveyed an impression of emerging scientific consensus "while industry and its partners ceded the science and fought on the economic issues." The document argued that this stance was miscalculated because a successful campaign to challenge the science "puts the United States in a stronger moral position and frees its negotiators from the need to make concessions as a defense against perceived selfish economic concerns." An auto industry public relations executive made a similar point: "Once you concede the science, all that is left is to argue the extent of liability and the timetable for cutting emissions. It's a lost cause." The importance of sustaining moral and intellectual leadership is clearly recognized here. The battle was lost, however; by late 1997, the business press in the United States and Europe was conveying the impression of scientific consensus (Raeburn 1997; Stipp 1997; *The Economist* 1997). The GCC was weakened by a series of defections: BP left in late 1997, Shell in 1998, and Ford in 1999. In early 2000, the GCC was reorganized to represent industry associations rather than individual member companies.

The growth of new organizations committed to a climate compromise undermined the GCC's claim to be the voice of industry on climate. Eileen Claussen, a former U.S. Assistant Secretary of State for Environmental Affairs and negotiator at the climate change negotiations, formed the Pew Center on Global Climate Change in April 1998. The Pew Center provides not only a channel of policy influence for member companies, but also a vehicle for legitimizing the new position. Thirteen prominent companies based in the United States, Europe, and Japan, including BP and Toyota, joined immediately and endorsed a series of newspaper advertisements stating that they "accept the views of most scientists that enough is known about the science and environmental impacts of climate change for us to take actions to address its consequences" (Cushman 1998). Other companies in sectors associated with low carbon technologies have increasingly exerted their collective voice. The Business Council for Sustainable Energy, for example, which has affiliates in the United States and Europe, represents insulation manufacturers and the fragmented renewable energy sector. Increasingly, however, it has attracted larger companies engaged in natural gas and electronic controls, including Honeywell, Enron, and Maytag.

Environmental activists and NGOs have played an active role trying to broaden the alliance of industries supportive of the Kyoto process. Of particular note have been the efforts of Jeremy Leggett, formerly of Greenpeace International, to gain the support of banking and insurance companies. Insurance companies have faced substantial claims for weather related property damage since the late 1980s, which some have attributed to the effects of climate change (Tucker 1997). Leggett has been instrumental in educating insurance companies about the potential risks and has successfully worked with some of them to take a more active role in the international climate negotiations (Leggett 2000).

These organizational realignments have been accompanied by the growth of the "win-win" discourse of "ecological modernization" (Hajer 1995) and a broader acceptance of the "precautionary principle." In a landmark speech on May 19, 1997, British Petroleum's Group Chief Executive John Browne stated that "the time to consider the policy dimensions of climate change is not when the link between greenhouse gases and climate change is conclusively proven, but when the possibility cannot be discounted and is taken seriously by the society of which we are part" (Castillo 1997). The need to reconcile economic strategy with this acknowledgement of the case for precautionary action makes "win-win" discourse very attractive. Ecological modernization puts its faith in the technological, organizational, and financial resources of the private sector, voluntary partnerships between government agencies and business, flexible market-based measures, and the application of environmental management techniques (Casten 1998; Hart 1997; Romm 1999; Schmidheiny 1992). The concept is reinforced by claims of significant cost savings from industry, such as BP's announcement in January 2003 that its success in reducing emissions by 10 percent (relative to 1990) had also generated $600 million in cost savings.

The "win-win" paradigm is a key discursive foundation for a broad coalition of actors supporting the emerging climate compromise. A number of industry associations, such as the Business Council for Sustainable Energy, the World Business Council for Sustainable Development, and the International Climate Change Partnership have adopted this language. Influential environmental NGOs in the United States, especially the World Resources Institute and Environmental Defense (Dudek 1996) have initiated partnerships with business to pursue profitable opportunities for emission reductions. Governmental agencies find "win-win" rhetoric attractive for reducing conflict in policymaking. In the United States, the joint EPA/Department of Energy Climate Wise program describes itself as "a unique partnership that can help you turn energy efficiency and environmental performance into a corporate asset" (U.S. Department of Education 1996).

These strategic and organizational shifts can also be related to convergent forces within the international oil and automobile industries. On the economic level, competitive pressure and interdependence compelled companies to respond to each other's moves. Toyota's commercial launch of the Prius, a hybrid electric-small gasoline engine car, in the Japanese market in 1998, took the industry somewhat by surprise. Honda leapfrogged Toyota and was the first to launch a hybrid in the U.S. market, in December 1999. Most American executives were dismissive of the prospects for the car in the United States, recalling that GM's electric vehicle had generated thousands of "preorders," which evaporated once the car was on the market in late 1995. Nevertheless, the U.S. auto companies were nervous that they might fall behind a competitor, and announced plans for their own hybrid vehicles for the mid 2000s. Ford's Kaericher remarked that:

of course we are concerned about what competitors are doing. We have to build a product that satisfies consumers and any insights into consumer demands are a scarce and valuable commodity. Maybe we have missed something.

Daimler's investment in April 1997 in the Canadian fuel cell company Ballard had a similar effect, with Ford joining the venture in December 1997. After Daimler announced a target date of 2004 for introducing a commercial fuel cell vehicle, Ford, GM, BMW, and Honda followed with similar pronouncements. As Hesse expressed it,

we look very closely at our competitors with respect to the fuel cell technology. Once we started pushing the issue, we found a very strong trend with our competitors to step in and allocate resources and target money to chase us.

Organizational restructuring within the auto companies provided another source of institutional change. By the mid 1990s, top management in Detroit was becoming increasingly international in outlook and was perhaps more sensitive to the market and regulatory environments in Europe and elsewhere. Ford implemented its Ford 2000 project, which pushed toward the rationalization and integration of production and management worldwide, and GM began to move in a similar direction. Ford Europe, for example, became responsible for the Fiesta-size class of small cars worldwide. According to one Ford Europe manager, "now we have global managers in top management, people who grew up in other cultures. Ford understands the importance of Europe now, and this really puts pressure internally on the US."

In the oil industry, even more than for automobiles, companies have broadly similar profiles, have access to global markets and resources, and have followed each other through the waves of diversification, restructuring, and consolidation. These structural similarities provide the capacity and incentives for companies to pursue similar strategies. The move from geographic organization toward globally integrated business units, increasingly based in subsidiary locations, reduces the impact of the home country environment on strategy making. The internationalization of senior management, which also reduces the institutional dominance of the home country, has proceeded slowly in the oil companies, although some change occurred between 1995 and 2000. Annual reports indicate that an increasing number of senior executives have spent significant portions of their careers outside the home country. Texaco managers explicitly associated the change in the company's position on climate with the appointment of Peter Bijur as CEO in 1996 and his openness to European perspectives.

Oil companies are also very sensitive to each other's actions. The 1989 Exxon Valdez oil spill, for example, stimulated concern amongst those competitors following similar procedures and routes, and constituted a catalyst for change in BP (Reinhardt 2000). In turn, John Browne's 1997 speech caused other companies to reconsider their positions. One Texaco executive stated that "Texaco has always been stronger in engineering than public relations, but we're trying to change. We saw how much mileage BP got from Browne's speech" (Castillo 1997). Texaco also began inventorying greenhouse gas emissions in 1998 after examining protocols used by BP and Shell.

In light of these convergent pressures, it is not surprising that some observers see more similarities than differences between the European and U.S.-based companies. The companies pursue similar strategies in their core oil and gas businesses, which account for the majority of revenues and profits. Investments in renewables could be seen as small, niche ventures in related industries rather than a major redirection of corporate resources. Exxon's recalcitrant position can perhaps be explained in terms of idiosyncratic firm-specific factors. A highly regarded internal scientist has played a leading role in the company's climate strategy, the company's tightly centralized structure has allowed for few dissenting voices, and its strong financial position provides no pressure for change. Texaco, by contrast, felt compelled to reevaluate its strategy as oil prices fell below $15 a barrel at the end of the 1990s. Recently, however, even Exxon appeared to be softening its stance; its 2000 Health Safety and Environment Report acknowledged that scientific evidence warranted some precautionary action to curtail greenhouse gas emissions.

The companies expressed only minor differences concerning the future of the oil industry. All expected oil production to peak around 2020 to 2030, with a slow subsequent decline, though Shell expected the peak toward the earlier end of the range. None of the companies expected renewables to pose major threats to oil and gas before midcentury due to cost and infrastructure limitations. Although the companies had all initially perceived climate change to be a serious business threat, their perception of interests had changed over time. Shell and Exxon concurred that the outlook was strong in the medium term; demand for gas for power generation was booming even without carbon controls. Oil was used primarily for transportation, and any improvements in fuel efficiency would be more than offset by growth in vehicle sales and miles traveled, particularly in developing countries. Air transportation was growing rapidly. Fuel cells for vehicles still faced many cost and technical barriers, and even if commercially successful, would utilize hydrogen derived from oil-based hydrocarbons in the medium term.

Participation in industry associations and institutional fora specific to climate change provide arenas within which expectations and understandings tend to converge. Key managers responsible for climate strategy in each of the oil companies studied were on first name terms and had met each other frequently during many official negotiating sessions and in other conferences and meetings. European companies have participated in the American Petroleum Institute and the GCC, and American companies attend European industry meetings. At the 1998

World Economic Forum meeting in Davos, a special session on climate change was organized for oil executives, after which John Browne said that he was "struck by the openness of the debate among senior people in the industry and in particular by the strong support for action expressed by Shell's chief executive Cor Herkströter and Texaco's CEO Peter Bijur" (Browne 1998). The London-based International Petroleum Industry Environmental Conservation Association (IPIECA), in which all the major oil companies participate, has an active working group on climate change. IPIECA has served as a particularly important site for the oil companies to discuss their views and reconcile differences, and IPIECA staff described a clear convergence trend. A July 1998 workshop on implementation mechanisms produced a stalemate, with U.S.-based companies concerned that any agreement to mechanisms could lead toward a binding treaty. By 2000, IPIECA was able to produce a document representing a common approach to mechanisms.

This common institutional environment provides a channel for the generation of consensus on the scientific and business threat posed by climate change, likely regulatory responses, and potential industry action in the political, market, and technological spheres. More broadly, the intense intercompany interactions promote the diffusion of new conceptual frames for considering the business-environment relationship. U.S.-based companies have moved toward accepting the need for some precautionary action, even in the absence of definitive scientific evidence.

Conclusions: Corporate Interests and the Emerging Climate Regime

It is somewhat ironic that the United States withdrew from the Kyoto Protocol in 2001 just as major sectors of U.S. industry were shifting toward a more accommodative position on emission controls. The withdrawal can be attributed to the ideological leanings of the incoming Bush administration, and to its close ties to the U.S. oil industry in particular. Recalcitrance on the part of the United States has not, however, been sufficient to stall the momentum of the emerging regime to control greenhouse gas emissions. Although it is increasingly unlikely that the 1997 Kyoto Protocol will enter into force as a formal mandatory regime, it is being supplanted by an informal regime of decentralized efforts at regional, national, and sub-national levels, involving multiple linkages and various forms of commitment from business, governmental authorities, and NGOs. While many expected the process to fall apart without U.S. participation, a Gramscian analysis suggests that the emerging climate regime can be understood in terms of a hegemonic bloc in the process of coalescing, as a diverse group of actors succeeds in coordinating their interests sufficiently through common discursive articulations and economic arrangements. This complex dynamic system has developed considerable coherence and momentum, and is thus able to continue its evolution even in the absence of single, albeit powerful, state.

At COP-6 in Bonn in July 2001, 178 countries agreed to move ahead with implementation of the Kyoto Protocol without U.S. participation. At the time of writing in November 2003, the fate of the Kyoto Protocol was in the hands of Russia, whose entry would signal that countries representing more than 55 percent of the emissions of industrialized countries had ratified the Protocol, thus giving it formal legal force. The specific mechanisms and targets agreed by the parties to the Protocol made it easier to bring reluctant countries on board and reduced the opposition by some industry sectors. Some features of the Protocol itself provide mechanisms that reduce the potential costs to companies and even offer potential profit opportunities. The main elements of the Protocol include mandatory but modest emission targets, which are substantially weakened by broad and flexible mechanisms for implementation and by weak enforcement. The inclusion of carbon sinks introduces considerable uncertainty and room for creative accounting due to the difficulty in establishing baselines and measuring changes. The ability to buy carbon credits in international emission trading schemes, which the EU had tried to limit, enables countries of the former Soviet Union to sell large amounts of "hot air" credits that they have available due to the collapse of their industrial base since 1990. This greatly diminishes the need for buyers of carbon credits to reduce domestic emissions from industry, transportation, or power generation. The Clean Development Mechanism and Joint Implementation reduce the adjustment burden and create profit opportunities for firms selling low-emission technologies.

Even if the international treaty does enter into force, many argue that it is fast becoming irrelevant, and that the more significant regime structures are growing organically from the initiatives of NGOs, companies, and authorities at multiple levels (Lee 2003). Nearly half the states in the United States are addressing climate change in some manner; many are drafting climate change action plans and enacting renewable portfolio standards, which require a growing percentage of generation to be from renewable sources. Ten northeastern states are designing an ambitious regional carbon cap-and-trade system for power generators, and California's legislature in 2002 began the process of regulating carbon emissions from automobiles; New York announced its intention to follow suit. The European Union has been preparing a regional carbon-trading system, due to start in 2005 and affecting the power, iron, steel glass, cement, ceramic, pulp, and paper industries. Several countries, including the United Kingdom, have been planning national energy and carbon taxes.

There are also many initiatives underway in the private sector, as companies realize the inevitability of the need to adjust to a carbon-constrained world. The U.S. withdrawal from Kyoto does not spare American multinationals, with production and sales across the globe, from the need to develop products and technologies to meet European regulations, and some proposed state-level requirements in the United States. Many companies, often in partnership with environmental NGOs, have begun to inventory their emissions and seek ways to reduce them. More ambitiously, several private initiatives have been established to create carbon trading systems among participating companies. The World Bank Prototype Carbon Fund (PCF) was established in 2000 as a "public-private partnership" between a few national governments, including the Netherlands, Sweden, Japan, and Canada, and twenty-six companies, including Hydro Quebec, Daimler-Chrysler, Shell-Canada, BP-Amoco, and numerous Japanese firms. The Fund's purpose is to raise $140 million for investments in renewables and efficiency in developing countries, projects that will earn carbon credits for the investing companies. Most recently, the Chicago Climate Exchange opened for business in October 2003 with twenty-two members, including American Electric Power and Ford. The members have committed to reducing emissions from North American operations by 1 percent a year for four years, and can engage in trading to meet those commitments.

While the momentum of this informal, fragmented regime is clearly gathering pace, and encompasses many diverse social and economic sectors, the effectiveness of these efforts at reducing emissions remains to be seen. Many countries are unlikely to meet their Kyoto targets, and it is unclear what all the voluntary efforts add up to in the absence of a global cap and meaningful enforcement. The price of accommodation with industry is that environmental goals are compromised to some extent. Plans for energy taxes in the UK, for example, have been substantially diluted to placate objections from various sectors, and the country will now probably miss its reduction commitments. Initial trades on the Chicago Climate Exchange have been priced very cheaply, at just under $1 per ton of CO_2, suggesting that the cap is not very stringent. The United States provided an explicit assurance that industry interests would be integrated into the regime at the negotiations in Geneva in July 1996, when the United States first agreed to a binding international agreement. Chief negotiator Tim Wirth promised that the United States would pursue "market-based solutions that are flexible and cost-effective," and that "meeting this challenge requires that the genius of the private sector be brought to bear on the challenge of developing the technologies that are necessary to ensure our long term environmental and economic prosperity" (Wirth 1996). Despite the formal withdrawal of the United States from the treaty, it appears that this goal has been achieved.

Note

1. All quotes not otherwise cited are from personal interviews.

References

Accu-Weather (1994). Changing weather: Facts and fallacies about climate change and weather extremes. State College, PA: Accu-Weather, Inc.

Anderson, P., & Tushman, M. L. (1990). Technological discontinuities and dominant designs: A cyclical model of technological change. *Administrative Science Quarterly*, 35, 604–633.

Bradsher, K. (1998, April 26). European auto division calling for improved fuel economy. *New York Times*, p. 24.

Bradsher, K. (1999, February 14). Making tons of money, and Fords, too. *New York Times*, p. 14.

Browne, J. (1998). Energy companies and the environment can coexist. *USA Today Magazine* (September 1).

Callon, M. (1998). *The laws of the markets*. Oxford: Blackwell.

Castillo, C. (1997). British Petroleum CEO Browne says firm will respond on global warming. *Stanford Report* (May 21). Available at <http://news-service.stanford.edu/news/1997/may21/bp.html>.

Casten, T. R. (1998). *Turning off the heat*. Amherst, NY: Prometheus Books.

Chen, M. J., & MacMillan, I. C. (1992). Non-response and delayed response to competitive moves: The roles of competitor dependence and action irreversibility. *Academy of Management Journal, 35,* 539–570.

Chen, M. J., & Miller, D. (1994). Competitive attack, retaliation and performance: An expectancy-valence framework. *Strategic Management Journal, 15,* 85–102.

Christensen, C. M. (1997). *The innovator's dilemma: When new technologies cause great firms to fail*. Boston: Harvard Business School Press.

Coen, D. (1999). The impact of U.S. lobbying practice on the European business-government relationship. *California Management Review, 41*(4), 27–44.

Corporate Europe Observatory (1997). *The weather gods: How industry blocks progress at Kyoto Climate Summit*. Amsterdam: Corporate Europe Observatory.

Corzine, R. (2000, November 2). Clock ticks down as Shell mulls break with tradition. *Financial Times*, p. 30.

Cushman, J. H. J. (1998, May 8). New policy center seeks to steer the debate on climate change. *New York Times*, p. 13.

Davis, R. E. (1996). *Global warming and extreme weather: Fact vs. fiction*. Washington, D.C.: Global Climate Coalition.

Dudek, D. J. (1996). *Emission budgets: Creating rewards, lowering costs and ensuring results*. New York: Environmental Defense Fund.

Eaton, R. J. (1997, July 17). Global warming: Industry's response. *Washington Post*, p. A19.

The Economist (1992, May 9). Europe's industries play dirty, pp. 91–92.

The Economist (1997, October 11). Sharing the greenhouse. *The Economist*, p. 20.

The Economist (2001, February 8). The slumbering giants awake. Energy Survey. *The Economist*, pp. 6–11.

Exxon (1999). *The Kyoto Agreement* <http://www.exxon.com/exxoncorp/news/publications/global_climate_change/globe3.html>.

Finger, M. (1994). Environmental NGOs in the UNCED process. In T. Princen & M. Finger (Eds.), *Environmental NGOs in world politics* (pp. 186–213). New York: Routledge.

Friends of the Earth International (1997). *Lobbying for lethargy: The fossil fuel lobby and climate change negotiations.* London: Friends of the Earth International.

Gelbspan, R. (1997). *The heat is on.* Reading, MA: Addison Wesley.

Greenpeace International (1997). *Oiling the machine: Fossil fuel dollars funneled into the U.S. political process.* London: Greenpeace International.

Haas, P. M. (1993). Epistemic communities and the dynamics of international environmental co-operation. In V. Rittberger (Ed.), *Regime theory and international relations* (pp. 168–201). Oxford: Clarendon Press.

Hajer, M. A. (1995). *The politics of environmental discourse: Ecological modernization and the policy process.* Oxford: Clarendon Press.

Hamilton, K. (1998). *The oil industry and climate change.* Amsterdam: Greenpeace International.

Hart, S. L. (1997). Beyond greening: Strategies for a sustainable world. *Harvard Business Review, 75,* 66–76.

Henderson, R., & Clark, K. B. (1990). Architectural innovation: The reconfiguration of existing product technologies and the failure of established firms. *Administrative Science Quarterly, 35,* 9–30.

Hirst, P., & Thompson, G. (1996). *Globalization in question.* Malden, MA: Blackwell.

Hoffman, P. (1999, April). DaimlerChrysler unveils liquid-Hydrogen NECAR 4 in U.S., Reaffirms 2004 launch date. *Hydrogen and Fuel Cell Letter* (Rhinecliff, NY).

Jasanoff, S. S. (1987). Contested boundaries in policy-relevant science. *Social Studies of Science, 17,* 195–230.

Kempton, W., & Craig, P. P. (1993). European perspectives on global climate change. *Environment, 35*(3), 17–45.

Knickerbocker, F. T. (1973). *Oligopolistic reaction and multinational enterprise.* DBA Thesis, Harvard Business School, Division of Research.

KPMG Environmental Consulting (1999). *KPMG International survey of environmental reporting 1999.* De Meern, The Netherlands: Institute for Environmental Management, University of Amsterdam, and KPMG.

Lee, J. (2003, October 29). The warming is global but the legislating, in the U.S., is all local. *New York Times,* p. 20.

Leggett, J. (2000). *The carbon war: Dispatches from the end of the oil century.* London: Penguin Books.

Levy, D. L., & Egan, D. (1998). Capital contests: National and transnational channels of corporate influence on the climate change negotiations. *Politics and Society, 26*(3), 337–361.

Levy, D. L., & Rothenberg, S. (2002). Heterogeneity and change in environmental strategy: technological and political responses to climate change in the automobile industry. In A. Hoffman & M. Ventresca (Eds.), *Organizations, policy and the natural environment: Institutional and strategic perspectives* (pp. 173–193). Stanford, CA: Stanford University Press.

Lippert, J. (1997, April 14). New R&D policy powers GM engine quest. *Automotive News*, p. 20.

Mansley, M. (1995). *Long term financial risks to the carbon fuel industry from climate change.* London: The Delphi Group.

Mazey, S., & Richardson, J. (Eds.) (1993). *Lobbying in the European community.* Oxford: Oxford University Press.

McLaughlin, A., Jordan, G., & Maloney, W. A. (1993). Corporate lobbying in the European community. *Journal of Common Market Studies, 31*(2), 191–212.

Montgomery, D. W. C. R. A. (1995). Toward an economically rational response to the Berlin Mandate. Washington, DC: Charles River Associates.

National Environmental Trust (1998). *Climate bulletin: Big oil's secret plan to block the global warming treaty* (Press Release www.envirotrust.com/rel98apr27.html). Washington, DC: National Environmental Trust.

Nauss, D. W. (1999, January 28). Auto makers are finding it's not easy being green. *Los Angeles Times.*

Newell, P., & Paterson, M. (1998). A climate for business: Global warming, the state and capital. *Review of International Political Economy, 5*(4), 679–703.

Oliver, C. (1997). The influence of institutional and task environment relationships on organizational performance: The Canadian construction industry. *Journal of Management Studies, 34*(1), 99–124.

Ozone Action (1996). *Distorting the debate: A case study of corporate greenwashing.* Washington, DC: Ozone Action.

Raeburn, P. (1997, November 3). Global warming: Is there still room for doubt? *Businessweek*, p. 158.

Reinhardt, F. (2000). Global climate change and BP Amoco. *Harvard Business School Case Study.* Case no. 700-106. Boston, MA: Harvard Business School Press.

Romm, J. R. (1999). *Cool companies: How the best businesses boost profits and productivity by cutting greenhouse gas emissions.* Washington, DC: Island Press.

Rowlands, I. H. (2000). Beauty and the beast? BP's and Exxon's positions on global climate change. *Environment and Planning, 18*, 339–354.

Schmidheiny, S. (1992). *Changing course.* Cambridge, MA: MIT Press.

Scott, W. R., & Meyer, J. W. (Eds.) (1994). *Institutional environments and organizations.* Thousand Oaks, CA: Sage.

104   David L. Levy

Sethi, S. P., & Elango, B. (1999). The influence of "country of origin" on multinational corporation global strategy: A conceptual framework. *Journal of International Management, 5*, 285–298.

Shnayerson, M. (1996). *The car that could.* New York: Random House.

Smircich, L., & Stubbart, C. (1985). Strategic management in an enacted world. *Academy of Management Review, 10*(4), 724–736.

Stipp, D. (1997, December 8). Science says the heat is on. *Fortune Magazine,* p. 126.

Stoffer, H. (1997, April 7). Projected fleet mpg drops; lowest since '80. *Automotive News,* p. 10.

Tucker, M. (1997). Climate change and the insurance industry: The cost of increased risk and the impetus for action. *Ecological Economics, 22*, 85–96.

United Nations Transnational Corporations and Management Division (1993). *Environmental management in transnational corporations: Report of the Benchmark Corporate Environmental Survey* (E.94.II.A.2). New York: United Nations.

U.S. Department of Education (1996). *Climate Wise DOE/EE-0071, EPA 230-K-95-003.* Washington, DC: Author.

US Newswire, (1999, February 11). ABB, Entergy, Shell International join growing corporate effort to address climate change. *US Newswire.*

Vogel, D. (1978). Why businessmen distrust their state: The political consciousness of American corporate executives. *British Journal of Political Science, 8,* 45–78.

Wapner, P. (1995). Politics beyond the state: Environmental activism and world civic politics. *World Politics, 47*, 311–340.

WEFA Group and H. Zinder & Associates (1996). *A review of the economic impacts of AOSIS-type proposals to limit carbon dioxide emissions (prepared for Global Climate Coalition).* Eddystone, PA: WEFA Group.

Wirth, T. E. (1996). Statement by Timothy E. Wirth, Under Secretary for Global Affairs, on behalf of the United States of America, at Convention on Climate Change, Second Conference of the Parties, July 17. Geneva, Switzerland: United States Mission, Office of Public Affairs.

World Economic Forum. (2000). Press release, January 27. Available at <http://www.weforum.org/site/knowledgenavigator.nsf/Content/Business%20leaders%20say%20climate%20change%20is%20our%20greatest%20challenge_2000?open&topic_id=>.

5

The Business of Ozone Layer Protection: Corporate Power in Regime Evolution

Robert Falkner

The Montreal Protocol of 1987 is frequently cited as one of the clearest examples of business influence over international environmental negotiations. Leading chemical firms, such as DuPont and ICI, played an important role in shaping governmental negotiation positions. Having initially opposed, and later supported, an internationally binding ozone regime, the chemical industry helped to make the Montreal Protocol a success. Subsequently, the producers of ozone-depleting substances (ODS), such as chlorofluorocarbons (CFCs), introduced substitute chemicals and quickly entered into a global race to capture the emerging market for ODS-free technologies and products. The relative success of the Montreal Protocol in reversing the trend towards ozone layer depletion, and the constructive role played by leading chemical firms, has given rise to the now widespread perception that proregulatory business interests and corporate involvement in international negotiations and implementation are key ingredients in effective international environmental governance. Put in a wider context, the experience of the ozone treaty suggests that global corporations are not simply part of the problem, but also part of the solution in international environmental protection. Understanding the dynamics of global environmental politics therefore requires a closer look at the political economy of state-firm relations in international regime creation. The intriguing question that the Montreal Protocol negotiations pose—and that is at the center of many contributions to this volume—is whether corporations have come to exert a pervasive, even hegemonic, form of influence in this global policy area.

This chapter examines the role of the corporate sector in international ozone politics and the sources of its political power. The process of corporate lobbying from the outset of the ozone controversy in the 1970s is well documented, and I will not cover it here (see Kauffman 1997; Levy 1997; Litfin 1994; Maxwell and Weiner 1993; Oye and Maxwell 1995). Let me therefore state at the outset in what ways the account presented here differs from previous studies.

First, this chapter shifts the focus from the pre-1987 negotiations to the treaty revision phase from 1988 to 1995, and thus to the evolution and implementation of the ozone regime. Regime evolution is far from being a trivial component of effective international governance. The Montreal agreement of 1987 provided only a partial, and in many ways unsatisfactory solution to the problem of ozone layer depletion. It was developments following its adoption that helped transform the protocol into an effective, and increasingly comprehensive, instrument for eliminating, and not just limiting, ODS emissions. As will be argued below, business played a key role in this process.

Second, the subsequent analysis highlights the role not only of CFC-producing chemical firms but also of CFC-using industries, which assumed greater importance in the implementation phase. It was the CFC conversion strategies of user sectors, combined with the technological solutions provided by the CFC producers, that shaped the pattern and speed of the CFC phase-out schedule adopted by the contracting parties.

Third, this chapter conceptualises corporate power as "technological power," thereby highlighting the critical role that corporate actors play in shaping the knowledge framework of international environmental policymaking. Technological knowledge and innovation were pivotal factors that drove the treaty revision process. Of course, new scientific findings that hardened the link between CFC emissions and ozone layer thinning played their role, too. But, ultimately, substitute technologies had to be found and commercially introduced if the ozone regime was to make a difference, and it was in this area that corporate actors set the parameters for political action. More than many other environmental issues, the problem of ozone layer depletion lent itself to a primarily technological solution, which helped to move the role of corporate actors centre stage.

The focus on technological power serves to challenge conventional perspectives on international environmental politics, which see international regulation as a key driving force in technological innovation. The state-centric tradition in international regime analysis, for example, tends to view regulations as "technology-forcing" and environmental protection as a top-down process of standard-setting and corporate adaptation. There is, of course, an element of truth in this. In many ways, the history of the Montreal Protocol provides a showcase for such state-led efforts to bring about a more environmentally sustainable form of technological progress. What is missing from this account, however, is the way in which technological change itself, and the corporate strategies and decisions behind it, shape international rule-setting and the regulatory process.

In contrast to the state-centric view on international regimes, this chapter argues that we need to take a closer look at the dynamics of state-firm relations in order to understand the sources of effective environmental governance. Firms are not simply rule-takers, but are intimately involved in regime creation and evolution. Their privileged position in directing technological innovation gives them special leverage in international politics. Technological power allows corporations to play a decisive role in shaping regulatory discourses, particularly with regard to the design and phasing of environmental regulations. The subsequent analysis looks at the way in which the strengthening of the Montreal Protocol post-1987 was influenced by technological and commercial decisions taken by the corporate sector. It provides a corrective to the view that treaty revisions follow a primarily state-centric logic that is informed by the accommodation of state interests and the accumulation of scientific knowledge.

The first section of this chapter provides an overview of the diverse corporate responses to the Montreal Protocol, focusing on the technological decisions taken by CFC producers and users. The second section examines the ways in which corporate strategy and technological innovation fed into the international political process that produced a series of treaty revisions between 1988 and 1995. The concluding section summarises the argument and draws out broader lessons for the study of corporate power in international environmental politics.

Corporate Responses to the Montreal Protocol

Chlorofluorocarbons, which were widely used in refrigeration, air conditioning, foams, aerosol and electronics production, were first linked to ozone layer depletion in 1974. The CFC-ozone depletion link started out as a highly contested scientific hypothesis, but scientific evidence in support of it grew over time. By the time atmospheric scientists publicised the discovery of the so-called "ozone hole" over Antarctica in 1985, environmentalists, policy-makers and even some industrialists had accepted the need to take precautionary action. Initially, though, the CFC producer and user industries opposed the scientific hypothesis and sought to discredit its scientific basis. The chemical industry followed the tried and tested strategy of denial and resistance, fighting the growing demands for regulation at national and international level (Roan 1989).

The first signs of a major strategic shift among industry leaders emerged in the mid-1980s, in response to growing scientific evidence of a link between CFCs and ozone layer depletion. The American chemical giant DuPont took the lead among CFC producers and declared in 1986 that it supported international regulations that would cap the future growth of CFC emissions. In doing so, the company single-handedly destroyed the hitherto united corporate front against international CFC restrictions. Many observers at the time felt that it had gambled on its technological strength in developing substitute chemicals that would gradually replace CFCs. By gaining a first mover advantage and promoting international regulations, DuPont was suspected of flexing its economic muscles in the political realm of negotiations in order to create and capture a new market for substitute chemicals. Indeed, the successful conclusion of the Montreal Protocol negotiations in September 1987 owed a great deal to the shift in corporate attitudes and the emerging consensus between governments, industry and NGOs on the need for international action. But the question that soon came to occupy policy-makers was whether CFC producers and users would cooperate in implementing the international agreement, and whether the CFC restrictions put in place could be strengthened in subsequent negotiations. In this, the corporate strategies of both CFC producers and users came to assume a more overtly political dimension.

The CFC Producers

Shortly after the Montreal Protocol had been signed in September 1987, the leading CFC producers declared their support for the groundbreaking international treaty. Some corporate representatives may have privately voiced reservations about the protocol's control measures, but the first official statements left no doubt that the chemical industry was willing to work with the new ozone treaty (*European Chemical News* 1987). This marked a shift in position for those firms, such as France's Atochem, which had opposed the treaty right until the end of the negotiations. They more or less grudgingly came around to the stance taken by DuPont, the world's largest CFC producer, in 1986, that an international regulatory framework was desirable.

As it turned out, industry's declarations of support were not a mere public relations exercise. The CFC producers could have tried to sabotage implementation of the Montreal Protocol at the national level, a strategy frequently employed by firms in other regulatory contexts. Instead, the chemical industry and some user industries took practical, and even innovative, steps to implement the ozone treaty. This had two important political consequences: first, corporate decisions on technological innovation played a central role in turning political will into economic reality. Second, the evolution of the ozone regime closely followed a technological, and thus largely corporate, logic. In other words, technological power in the hands of corporations both gave teeth to the international regime and shaped the regulatory discourse that underpinned its evolution.

The first evidence of industry's newfound cooperative spirit came in the form of an unprecedented move to pool the testing of CFC alternatives. In late 1987 and early 1988, two industry programs were created to assess the environmental acceptability (Alternative Fluorocarbons Environmental Acceptability Study—AFEAS) and toxicity (Program for Alternative Fluorocarbon Toxicology Testing—PAFT) of alternative chemical compounds. The industry cooperatives counted among their founding members thirteen CFC producers from around the world and had grown to seventeen by the mid-1990s. While these activities were restricted to the noncompetitive area of chemicals testing, some bilateral programs went as far as combining research and development efforts: Kali Chemie and ISC announced a joint effort to develop substitutes in the area of refrigeration and foam blowing; and Atochem and Allied-Signal set up a joint research programme covering a range of substitutes. As soon as the race to find CFC substitutes had started, it became clear that in order to succeed, chemical firms needed to have sizeable research budgets and global presence in the CFC business (*Chemical Engineering* 1988b; *European Chemical News* 1989).

Beneath the surface of industry cooperation, therefore, a fierce commercial battle for the substitutes market unfolded. In contrast to the smaller CFC producers, large chemical firms such as DuPont, ICI, or Atochem were already in a leading position. Not only did they have the necessary financial and organizational resources to invest in new technologies; they could also build on the experience of the 1970s, when the first ozone controversy prompted them to look into the potential for substituting CFCs. DuPont had spent some $70 million on substitutes research in the second half of the 1970s, and decided in 1986, before the Montreal Protocol was signed, to revive this program. After the signing of the Protocol, DuPont stepped up its expenditures and committed over $30 million in 1988 and over $45 million in 1989 to finding CFC alternatives—the largest of all research efforts undertaken in the late 1980s (*European Chemical News* 1989; *Manufacturing Chemist* 1988).

CFC producers played a key role in translating the political regulations of the Montreal Protocol into market signals, and vice versa. As policymakers looked to the chemical industry for technological solutions to ozone-layer depletion, the investment decisions of the major CFC producers assumed an important political dimension. They became the critical link between international regulation and changes in CFC production and consumption patterns, and thus the effectiveness of the ozone regime. The negotiating parties never seriously considered the possibility that governments themselves might set up research programs to find substitutes. It was therefore inevitable that the chemical industry came to shape political actors' perceptions of the technological feasibility of reducing, and eventually eliminating, CFC emissions.

One of the first technological decisions taken by the leading chemical firms concerned the type of chemical compound that would replace

CFCs. From the beginning of the ozone controversy, the chemical industry had placed its hopes on finding chemical substances that would be similar to CFCs but with a reduced impact on the ozone layer, so-called "drop-in" substitutes. Finding functionally identical, or at least similar, CFC replacements was no easy task, however. After all, the commercial success of the family of CFCs rested on their unique combination of nonflammability and low toxicity. Any other chemical substance would almost inevitably require some kind of trade-off: reducing the ozone-depletion potential could often be achieved only in exchange for higher toxicity or reduced flexibility in commercial applications. Because of this restriction, the chemical producers in the United States and Europe initially concentrated their efforts on close relatives to the widely used CFC-11 and CFC-12 (*IER* 1988).

The leading contenders to replace the most common CFCs were hydrogenated CFCs—most notably CFC-22, later renamed HCFC-22—and fluorocarbons without the ozone-depleting substance chlorine, so-called hydrofluorocarbons (HFCs). Due to their lower ozone-depleting potential, HCFCs were not included as regulated substances in the 1987 Montreal Protocol and therefore became a major component of the chemical industry's substitution strategy. HFCs, which were thought to pose no threat to stratospheric ozone, promised a more long-term solution for replacing CFCs, and were being promoted primarily in refrigeration and air-conditioning uses (e.g., HFC-134a). Despite continuing problems with toxicity and concerns about the contribution that HFCs made to global warming, a race soon unfolded to capture the emerging market for HFCs. In 1988, CFC producers were not expected to be able to produce HFC-134a on a commercial scale before 1993 (*Manufacturing Chemist* 1988). But only a year later, DuPont announced it was leading the race to develop a manufacturing process, and said it would begin commercial production by the end of 1990, several months before ICI was expected to bring its first HFC-134a facility on-stream (ENDS 1989). In similar fashion, the chemical firms rushed into expanding production of HCFC-22, despite warnings by scientists as early as 1988 that HCFC-22 may soon be considered an unacceptable substitute (*Manufacturing Chemist* 1988). Industry and government officials in North America and Europe were keen to see "drop-in" substitutes enter the market quickly, and thus largely ignored warnings that HCFCs and HFCs themselves might become the subject of future regulations.

While most CFC producers kept an open mind about potential substitute compounds, competitive pressures gave rise to at least two principal product strategies. The first group of producers, consisting of DuPont, Elf Atochem, and Montedison, simultaneously expanded production of HCFCs and developed varieties of HFCs. This strategy was endorsed by the Alliance for Responsible CFC Policy, a U.S.-based coalition of international CFC producers and users, which warned policy makers not to pursue hasty reduction schedules for these two types of transitional substances (Alliance for Responsible CFC Policy 1989). In contrast, ICI moved more decisively into HFC production. The company increased existing HCFC production without adding further production capacity to meet short-term increases in demand, but sought to convince its customers and the British government that an early switch to HFCs was feasible and the most desirable conversion strategy (Jordan 1997, 16–17). As a consequence of these two substitution strategies, U.S. and European positions on the future regulation of transitional substances began to diverge, with the latter beginning to argue for more forceful reduction targets for HCFCs later in the treaty revision process. The result of this was a near reversal of negotiation roles: whereas in the early to mid-1980s it was mainly the United States that argued for a comprehensive treaty to reduce CFC emissions, the Europeans, with the exception of France, were leading the campaign for an early phaseout of certain transitional substitute chemicals in the early 1990s.

The CFC Users

Unlike the chemical industry, most of the CFC user industries were not actively involved in the ozone controversy until after the Montreal Protocol was signed. And even then, many companies took their time to react to the CFC reduction program, assuming that either the CFC producers would develop alternative substances or policymakers would leave sufficient time for them to adjust. Only a minority of user firms took up the challenge and initiated ambitious efforts to eliminate CFC use. As will be shown below, the strategic choices made by the CFC user industries were to have an important impact on the evolution of the CFC phaseout regime.

The divergence in user industry approaches is striking, for there is little in the institutional design of the Montreal Protocol that can account for this.[1] Instead, we need to look at business strategies, market structures and corporate networks to find explanations for the variation in corporate responses, which, in turn, influenced the evolution of the international ozone regime. We can distinguish between three major factors that shaped the user industry response to the Montreal Protocol:

First, at a fundamental level, the nature of CFC usage influenced corporate strategies. In some cases, particularly in the aerosol industry, low technical barriers to substitution allowed for a relatively rapid conversion process. Second, the heterogeneity of the CFC user industries, ranging from small-scale refrigeration and air conditioning service units to large-scale electronics manufacturers, accounted for a certain degree of variation in corporate responses. Unlike the small group of large CFC producers, the user industries were too diverse to coordinate their activities and approach the conversion problem in a concerted manner. Third, market structures within these industry segments and corporate networks between producers and users also played an important role. For example, where the CFC producers enjoyed a close working relationship with CFC users, as in refrigeration, they maintained a strong influence over the choice of substitute technologies by their corporate customers, which, in turn, gave them greater leverage in the treaty revision process.

As the case of the CFC aerosol, solvent and refrigerant user industries shows, corporate decisions on technological innovation had an indirect impact on the path and speed of CFC conversion. This provides a corrective to state-centric perspectives on the technology/regulation nexus that tend to view international regimes as "technology-forcing." Once corporate agency is moved centre-stage, the impact of pathways of technological change on regime evolution becomes apparent. It is in this sense that corporate agency in directing technological innovation can be said to have a political dimension. In the context of the Montreal Protocol, technological innovation was essential to the effective implementation of the global CFC reduction plan, but could not be taken for granted. It depended, to a large extent, on the decisions taken by both CFC producers and users. Although an important factor in driving technological change, the Montreal Protocol's regulatory framework alone cannot explain the pattern and speed of CFC conversion.

The extent to which the corporate sector was able to shape regime evolution depended, inter alia, on its ability to unite behind a common strategy. Such corporate unity, however, was an elusive quantity in ozone politics. Differences in commercial interests and political strategies emerged not only between CFC producers and users, but also within the user industries. Where CFC user firms opted for the complete elimination of CFCs and substitute chemicals, as in the electronics sector, the position of the chemical industry, both in economic and political terms, suffered a serious blow. In contrast, where the user industries chose to rely on substitute chemicals provided by the CFC producers, as in the case of refrigeration and air conditioning manufacturers, the position of the chemical industry was strengthened in international ozone politics. The pervasive reality of business conflict thus served to curtail the political clout of the major CFC producers that had actively engaged with the international negotiations. The dynamics of competition and conflict in the business sector, rather than its structural power, became key determinants in the emerging relationship between the corporate sector and states in the post-1987 treaty revision process.

The aerosol industry was one of the first CFC users to react to the Montreal Protocol. In the case of CFC propellants, technical barriers did not stand in the way of conversion. This was evident from the US aerosol sector, which had phased out ozone-depleting propellants by the late 1970s. While most European aerosol firms had successfully resisted CFC restrictions in the 1970s, the Montreal Protocol forced them to reconsider their stance. Initially, the European aerosol industry hoped that the chemical industry would come up with "drop-in" alternatives. European CFC producers initially suggested HCFC-22 as the main substitute (ENDS 1987), which required only minimal process changes. However, within two years of agreement of the Montreal Protocol, a fundamental shift was underway in the aerosol market. Splits within the European aerosol industry had emerged. As soon as individual firms, such as Johnson Wax and Talbec, opted for non-ODS products that they were advertising as "ozone friendly," others felt they could no longer support HCFC-based solutions. The European aerosol industry switched en masse to hydrocarbons as the preferred alternative propellant. Although requiring higher initial investment costs to convert existing production plants, hydrocarbons turned out to be cheaper than HCFCs and had no negative impact on the ozone layer.

As a result of this market shift, the pattern of CFC propellant replacement in Europe closely followed the example set by the United States in the 1970s. As a consequence, the CFC and HCFC producers were left with no leverage over the aerosol industry and were forced to write off any chances of preserving the HCFC substitutes business in the aerosol sector. By the end of the 1980s, most European countries were well on course to meet the EC-wide target of 90 percent CFC reduction in the aerosol sector by the end of 1990 (UNEP 1989a, 13). Developments in the aerosol sector thus sent strong signals to policymakers in Europe in the run-up to the first treaty revision in 1990, enabling them to consider tougher CFC restrictions than had previously been envisaged.

The CFC solvent user industries reacted to the 1987 ozone treaty in the same reluctant and defiant manner as most other user industries. The electronics industry, the main user of CFC solvents such as CF-113 and methyl chloroform, had been arguing for some time that CFCs were essential to modern processes of electronics manufacturing and could not easily be replaced (Zurer 1988). Their position was strengthened when chemical firms reported that finding a substitute would prove far more difficult in the case of CFC-113 than CFC-11 and CFC-12 (*Chemical Engineering* 1988a). CFC-113 replacement was further complicated by the fact that the world's largest buyer of electronics goods, the U.S. military, stipulated the use of CFC solvents. Initially, therefore, the user industry expected other actors, especially the chemical industry, to take the lead in the search for alternatives. It also placed its hope on persuading governments to draw out the CFC-113 phaseout schedule, to allow for an economically painless conversion programme in a key industrial sector.

Three factors, however, brought about a change to the electronics industry's position, making it one of the first industries completely to eliminate ODS. First, CFC solvent usage during cleaning processes made up only a small fraction of the value of the final product. CFC usage formed part of the manufacturing process, not the end product. Second, the electronics industry did not consider its links to the chemical industry as essential to its business success. The chemical industry thus had limited influence over the CFC conversion process in this sector. Third, the success of the electronics industry was built on the ability to innovate and rapidly respond to changing market conditions and technological advances. This suggested that the industry was more inclined to consider other solutions for dealing with the CFC problem than those provided by the chemical industry.

The first efforts to find an alternative cleansing technology were made only shortly after the Montreal Protocol was adopted. AT&T and Petroferm announced in 1988 that a naturally derived product could be used to deflux electronic circuit assemblies, thus making CFC-113 potentially replaceable (EPA 1997, 68). Other firms developed similar cleansing methods that eliminated CFCs, or introduced process changes that made the cleansing of electric circuit boards redundant. Several multinational electronics firms, including AT&T and Nortel, investigated this substitution strategy, and within only a few years set up plans for the elimination of CFC-113 from electronics manufacturing.

As a consequence of these initiatives, virtually all major electronics companies committed themselves to eliminating CFC use by 1995, and many reached this goal much earlier. Some individual manufacturing plants had already in 1991 achieved 95 percent reduction of CFC usage below the 1987 level, and the overall pace of CFC-113 reduction was uniformly described as fast (Pollack 1991). At a time when the chemical industry was still searching for a replacement of the "magical" CFC-113, most electronics firms had already embarked on a process of eliminating all CFC solutions. The effect of these initiatives was dramatic. By 1992, worldwide CFC-113 consumption had fallen to 126,500 tonnes, down from 276,700 tonnes in 1988 (Makhijani and Gurney 1995, 172). Within a few years, one of the most intractable cases of CFC usage had nearly disappeared from the international agenda, against all expectations of policymakers and industrialists.

The refrigeration and air conditioning industry was one of the most reluctant user industries to respond to the Montreal Protocol. While other sectors managed to reduce their CFC consumption in the second half of the 1980s, the use of ODS in refrigeration and air conditioning went up, both as a proportion of overall global ODS consumption and in absolute terms, from approximately 420,000 tonnes in 1985 to over 480,000 tonnes in 1990. It was only in the 1990s that the sector began slowly to reduce consumption of these substances (Makhijani and Gurney 1995, 132–133).

The main reason cited by the industry for its belated reaction was technological barriers to replacing existing refrigerants. In the absence of a readily available "drop-in" substitute, CFCs were widely seen as "essential" to the proper functioning of residential and commercial cooling systems. However, on closer inspection, other factors related to market structures and corporate strategies also played an important role in shaping the industry's strategic response. The close relationship between the chemical industry and refrigeration and air conditioning manufacturers prevented a more radical redesign of cooling systems which would dispense with ODS altogether. Instead, the refrigeration and air conditioning industry dragged its feet over CFC replacement and relied on the chemical industry to come up with solutions: initially, ODS with low ozone depletion impact, and later HFCs, particularly HFC-134a.

The CFC producers first offered HCFC-22 as the optimal substitute for CFC-12 refrigerants, leading refrigeration manufacturers down a path that would later complicate the complete phaseout of ODS. Only by the mid-1990s, in response to regulatory restrictions on HCFCs, did the refrigeration industry introduce HFC solutions for both refrigerants and insulation (Somheil 1996, 29). Both these substitute choices were challenged by environmental campaign groups, who argued that an entirely different option—hydrocarbons—could replace existing technologies. Greenpeace, in particular, led an international effort to convince refrigeration manufacturers and consumers of the benefits of hydrocarbons as refrigerants, a technology that had already been developed but was rejected by manufacturers. Eventually, the Greenpeace campaign proved successful in a number of European countries, but failed to have an impact on the North American market.

Greenpeace's effort to introduce a CFC-free refrigerator began in late 1991 in Germany and led to an agreement in June 1992 with DKK Scharfenstein, a near-bankrupt East German manufacturer, to produce ten CFC-free refrigerators in a pilot project (Ayres and French 1996). Greenpeace subsequently used its campaigning clout to help market the new CFC-free refrigeration technology, dubbed "Greenfreeze," in Germany and abroad. Germany's main refrigerant manufacturers initially opposed the campaign, but following a shift in the market and the public relations damage suffered by their anti-Greenfreeze campaign, quickly adopted the new technology. Within three years, hydrocarbon systems were used in 90 percent of the German household refrigeration market by 1996. The decision to replace HFC-134a with hydrocarbons dealt a major blow to the German chemical producer Hoechst and its efforts to expand HFC production as part of its substitution strategy (*IER* 1993e).

Outside Germany, only a small number of countries adopted the hydrocarbon technology, among them Switzerland and Nordic countries. In North America, the new technology failed to make an impact with domestic users. This was to a large extent due to U.S. health and safety regulations, which effectively prevented hydrocarbon refrigerators from entering the market, and the perceived lower energy efficiency of the European competitors' model. Moreover, the U.S. refrigeration industry had come up with what it advertised as being "CFC-free" using HFC-134a as refrigerant and HCFCs for foam insulation (Cook 1996, 6). The U.S. industry had no incentive to reverse its technological choice and blocked Greenfreeze technology from advancing further in the North American market.

In the end, the chemical industry's HFC-based strategy paid off. Hydrocarbon systems succeeded, albeit mainly in European markets, and even there only in small-scale household refrigeration units. Large-scale commercial refrigeration systems, which are by far the largest market segment, continued to rely heavily on the chemical industry's preferred substitutes, HCFCs and HFCs. Although the CFC producers failed to achieve exclusive dominance in the refrigerant substitutes market, they nevertheless secured the larger part of the substitutes market and effected a fundamental challenge to their dominant position in the refrigeration substitute business.

The experience of the refrigeration industry suggests that despite regulatory pressure and available substitutes, corporate actors exercise significant influence over the path of technological innovation and the speed of CFC reduction. The industry's reluctance to phase out CFCs and its reliance on HCFC and HFC substitutes played a major part in holding back the phaseout especially of the transitional substances, thus providing a boost to the political position of the CFC producers in the treaty revision process.

Corporate Power and the Treaty Revision Process

From CFC Reduction to Phaseout

As discussed previously, the chemical industry had signalled to negotiators its willingness to work with the Montreal Protocol. The CFC producers were confident that they could find substitutes for the two most important ODS, CFC-11 and CFC-12, but warned that finding substitutes for CFC-113 would prove more difficult (*Chemical Engineering* 1988a). Given that the Protocol mandated only a 50-percent CFC reduction over a ten-year period, governments and industry had reason to believe that implementing the ozone treaty was not an impossible feat. The important question was, however, whether the CFC restrictions could be further tightened in subsequent negotiations.

Corporate support for international regulation was more elusive when it came to environmentalists' demands for a complete phase out of CFC production. The major CFC producers made it clear that support for the Montreal Protocol was contingent on the adoption of a measured regulatory approach that would respect technical realities and commercial interests. In 1987, both CFC producers and user industries did not accept that a strengthening of the protocol, let alone a complete phaseout of CFCs, were economically and technically feasible. Developing CFC substitutes would take a considerable time, and even if some substitutes were to be found, not all CFC uses could be eliminated: so-called "essential uses" of CFCs were widely regarded to stand in the way of more stringent CFC restrictions.

More than ever before, perceptions of technological uncertainty came to dominate international ozone politics. The focus on perceived technical barriers to phasing out CFC emissions enhanced the role of corporations. As it turned out, corporate decisions, more than any political or scientific developments, brought about a radical change in the regulatory discourse. The key event occurred in March 1988, when DuPont broke ranks by announcing that it endorsed the target of a complete phaseout of CFC production. It was the second time that DuPont had taken the lead in ozone politics. The resulting rift in the international business coalition created political space for introducing tougher CFC regulations. DuPont's latest move, based on a unilateral strategic decision by the world's leading CFC producer, was to have a crucial impact not only on Washington's negotiating position. It also, and perhaps more importantly, played an important role in shifting the regulatory discourse towards the complete elimination of ODS.

DuPont took the decision shortly after NASA's Ozone Trends Panel published new findings on March 15, 1988, which raised serious questions about whether the Montreal Protocol's restrictions on CFCs were sufficient to protect the ozone layer. DuPont managers involved in the company's decision later attributed it to growing scientific evidence in support of further CFC restrictions—a view that supports epistemic community approaches that emphasise the role of growing scientific consensus in regime evolution (Glas 1988; Haas 1992).

However, DuPont's policy change was a politically significant step in its own right, not simply a reflection of the emergence of a scientific consensus. The NASA report did not present conclusive evidence in favor of the CFC-ozone loss theory. Nor did it mandate any particular policy response. DuPont and the other CFC producers could have insisted—as most of the user industries continued to do—that the Montreal Protocol's provisions represented a reasonable compromise in the interest of precaution, in light of the remaining uncertainties and economic costs of conversion. But, having taken the lead in ozone politics in 1986, DuPont saw the NASA report as signalling a trend in the scientific discourse that pointed in only one direction, toward a strengthening of the ozone regime. The logical conclusion was for DuPont to move ahead of the game and throw its weight behind a total phaseout goal: This, the company hoped, would make it easier to cooperate with policymakers in designing an "orderly" phaseout of CFCs. As the sequence of events following the Montreal Protocol demonstrated, this strategy largely paid off.[2]

Corporate decisions also played an important role in determining Europe's response to the Montreal Protocol. The CFC producing countries which had been most reluctant to agree to the ozone regime—Britain and France—continued to act as a brake on European decision making after 1987. Having agreed to CFC reductions in 1987, the EC took over a year to translate the international treaty into community law (Jachtenfuchs 1990). But within a relatively short period of time, Europe moved from a policy of foot-dragging to political leadership in speeding up the CFC phaseout. To be sure, domestic factors, particularly a strengthening of environmental campaigns across the continent, were an important factor. But given the closeness of industry-government links in Europe's corporatist environment, changes in corporate strategy played an equally important role. This was most prominently the case in Germany, which played a key role in overcoming British and French obstinacy within EC institutions.

In late 1987 and throughout 1988, Greenpeace and other campaign groups targeted Hoechst, Germany's biggest CFC manufacturer, as well as selected user industries in their ozone campaigns. Just as in the United States in the 1970s, the initial focus was on CFC use in aerosol manufacturing. As it turned out, Germany's aerosol industry was an easy target. It quickly broke ranks with other user industries and gave up its initial opposition to a complete CFC phaseout, having already agreed to a CFC reduction schedule in the run-up to the Montreal Protocol agreement. Other user industries, although reacting more slowly, followed suit and set the signals for a relatively early phase out of CFCs in Germany (*FAZ* 1988).

The CFC producers's response was more cautious, but was soon followed by a major shift in strategy. In the first few months after the signing of the Montreal Protocol, Hoechst demonstrated good will by setting up the first European recycling system for refrigeration liquids containing CFCs, while remaining critical of calls for more stringent CFC regulations (*Der Spiegel* 1988). It was in December 1988, in response to a parliamentary commission's call for a 95 percent CFC reduction, that Hoechst declared its support for an eventual phase out of CFCs. Hoechst was in a strong position to follow through this new strategy. By late 1988, the company's own reduction program was already three years ahead of the Montreal Protocol's second phase, and it was able to commit itself to a complete elimination of CFC production by 1999 (*SZ* 1988). The company eventually brought forward the phase-out date to 1994, becoming the first chemical company to stop CFC production (*IER* 1994b).

Having followed DuPont's leadership in 1988, Hoechst took the lead among its competitors in April 1989 when it went public with its new phaseout target of 1995, which was also adopted by Solvay, Germany's only other CFC producer.[3] The two firms urged their government to support the development of substitute chemicals and to work for a European-wide harmonisation of CFC reductions in order to create a level playing field. While Hoechst's move to some extent reflected the growth of the anti-CFC movement in Germany, pressure from environmental groups alone cannot explain Hoechst's strategic shift. After all, the decision to phase out CFC production had been taken well before Greenpeace's campaign reached its climax, and greatest public impact, in the summer of 1989, when activists climbed onto cranes at Hoechst's Frankfurt production facility (*FAZ* 1989). At the time, the company was going through a major strategic change which saw higher-profit specialty chemicals promoted at the cost of low-profit bulk chemicals such as CFCs. Given that only 0.5 percent out of a total annual turnover of around DM 40 billion in the late 1980s resulted from CFC production (*Der Stern* 1989), Hoechst managers were not willing to attract any more negative publicity in connection with the CFC-ozone controversy. To the dismay of Hoechst, however, the company found it difficult to capitalize on its leading position in the phase out of CFCs and continued to be the target of environmental campaigns for years to come.

Encouraged by these developments in Germany's CFC industry, the German government adopted a national CFC phaseout plan that aimed for a reduction of CFC usage of 95 percent by 1995, but stopped short of ruling out CFC production altogether. Germany then took this new policy to the EC-level and advocated a 1995 deadline for the whole community. This was unrealistic, but sent a strong signal to the recalcitrant member states. The European Commission itself had proposed a 1997 target, while France, Britain and Spain argued for 2000 to be kept as the phase out target. The EC eventually decided to aim for a global phase-out from 1997 onwards, and to ban the five regulated substances of the Montreal Protocol by 2000 (*IER* 1989). Germany had succeeded in nudging the EC position into a more proactive direction. As in the case of DuPont, a strategic shift by a leading European CFC producer had far-reaching political consequences.

Revising the Ozone Treaty: From Helsinki (1989) to London (1990)

The decision by the EC to support a complete phase-out by the end of the century, and to push for an earlier date if possible, had an important signal function for the upcoming First Meeting of the Parties in Helsinki in 1989. Only one day after the EC had decided on its new position, President Bush announced that the United States would also phase out CFCs by 2000, although he qualified the statement with the condition "provided that safe substitutes are available" (*IER* 1989). Although the Helsinki meeting produced only a nonbinding declaration, the EC and the United States were able to lay the ground for a major revision of the Montreal Protocol at the Second Meeting of the Parties in London in June 1990.

In the run up to the Helsinki conference, UNEP's assessment panels produced the first set of reports on the state of knowledge in the areas of atmospheric science, environmental impact of ozone depletion, and technological and economic aspects of CFC conversion. The Synthesis Report strongly suggested that the long stratospheric lifetime of CFCs made a wait-and-see approach undesirable. A complete and timely phase-out of all major ODS was, as the report put it, "of paramount importance in protecting the ozone layer" (UNEP 1989c: 28). Although this statement played an important role in strengthening the resolve of governments to revise the Montreal Protocol, it is important to note that by that time the major CFC producers had already committed themselves to an eventual phaseout of CFCs. This corporate commitment was echoed by the 1989 technological assessment panel, which stated that:

Based on the current state of technology, it is possible to phase down use of the five controlled CFCs by over 95 percent by the year 2000 . . . Given the rate of technological development, it is likely that additional technical options will be identified to facilitate the complete elimination of the controlled CFCs before the year 2000 (UNEP 1989b, ii).

The UNEP technology review thus confirmed what industry insiders had known for some time: that technological innovation could significantly reduce the time needed to complete the CFC substitution process. This did not mean that an international agreement on a revised CFC phase out schedule was now within easy reach. Far from it, negotiations on the path and timing of the CFC elimination programme proved to be complex. What had changed, however, was that scientists, environmentalists and leading industrialists had forged a consensus on the need to close down the CFC business.

What was still to be resolved was the exact phase-out schedule. In deciding this question, corporate decisions on technological change were of paramount importance. The contracting parties recognised this by inviting industry experts to join UNEP's Technology Assessment Panel, which was in a privileged position to shape the parameters of the regulatory discourse. To be sure, industry representatives could not impose their commercial interests on the international negotiations via authoritative technological assessments. Panel members were expected to act as technical experts and not as industry representatives. Moreover, the panels could only offer advice, whereas it was for the negotiating parties to reach a compromise on specific reduction schedules. Companies therefore continued to rely heavily on lobbying their governments in order to protect their particular commercial interests. And with CFC producers and users often pursuing divergent substitution strategies, the success of such lobbying efforts was far from guaranteed.

From a corporate perspective, the critical issues on the agenda of the Second Meeting of the Parties in London in June 1990 were (1) the inclusion of other ODS, such as HCFCs, in the list of regulated substances; and (2) the tightening of the existing CFC reduction schedule. In principle, all chemical producers were keen to safeguard their investment in HCFCs for as long as possible. They argued that if HCFCs were banned in the near future, investment in their production would be at risk and user industries would be reluctant to cut back on their CFC usage until a safe and acceptable long-term alternative had been found. Governments on both sides of the Atlantic were largely sympathetic to these arguments. In early 1989, the U.S. industry had been given assurances by EPA that the U.S. government would seek to protect the use of HCFC-22 at the forthcoming Helsinki negotiations (*ACHRN* 1989). And the EC and Japan were even more determined at that time to guarantee the long-term availability of HCFCs.

Eventually, the position of HCFC producing countries prevailed over the calls by those countries that favored an early phaseout of the transitional substances. The final text of the agreement called only for the use of HCFCs to be limited "to those applications where other more environmentally suitable alternative substances or technologies are not available," and to be ended "no later than 2040 and, if possible, no later than 2020" (quoted in Benedick 1991, 263–64). This voluntary phaseout date was heavily criticised by environmentalists who pledged to push for HCFCs to be formally brought into the regulatory regime at the next Meeting of Parties in 1992. They were concerned about the contribution of HCFCs not only to ozone layer depletion but also to global warming.

On the question of revising the existing CFC reduction schedule, it was the EC that favoured a stricter approach than that proposed by the United States and Japan. The EC had put forward the year 1997 as the final deadline for eliminating CFC use, while the United States American and Japan preferred the year 2000. The difference in negotiating positions reflected, to a large extent, differences in corporate interests. While Germany's Hoechst had declared itself ready for an earlier phaseout date, thereby undermining the more cautious approach adopted by ICI and Elf Atochem, the U.S. producer DuPont remained doubtful about a complete phaseout in 1997 (Benedick 1991, 171–72). In the end, the lowest common denominator position prevailed, and a phased elimination programme for the five main CFCs was agreed with the year 2000 as the final phaseout date.

Among the other outcomes of the conference, the EC achieved a concession that allowed its CFC producers to rationalize production EC-wide. Environmentalists scored a victory by having methyl chloroform included as a regulated substance, against chemical industry lobbying at the conference. Reducing methyl chloroform usage (primarily as a solvent) promised the single most important short-run contribution to lowering stratospheric ozone concentrations (Litfin 1994, 151). Moreover, the progress made by electronics firms in replacing ODS in cleaning processes undermined the lobbying effort by methyl chloroform producers. UNEP's technology panel had concluded that 90 to 95 percent of ODS use as solvents could be eliminated by the year 2000 (Litfin 1994, 151).

Moving toward Phaseout: Copenhagen (1992), Vienna (1995)

As Karen Litfin (1994, 156) points out in her book *Ozone Discourses*, "two primary factors . . . drove the treaty revisions, the scientific observations of unprecedented ozone losses and the rapid progress in generating alternative technologies." In the aftermath of the 1990 conference, new scientific studies painted an even bleaker picture of the environmental damage that was being done to the stratosphere. At the same time, the second report of UNEP's technology and economic assessment panels in 1991 documented the rapid progress in finding CFC substitutes that had been made since the Montreal Protocol was signed. The report predicted that by 1992, CFC consumption would be reduced to 50 percent of 1986 levels, a target that the 1990 revisions had set for the year 1995. Furthermore, the report suggested that virtually all consumption of CFCs could be eliminated between 1995 and 1997 (Litfin 1994, 164). Given this optimistic outlook on the implementation of the Montreal Protocol, the next revision of the treaty, at the Fourth Meeting of the Parties in Copenhagen in November 1992, was widely expected to produce a further tightening of the CFC restrictions.

Indeed, one of the major outcomes of the Copenhagen conference was a relatively uncontroversial agreement on revised phase-out deadlines for the main ODS. Most CFCs were to be eliminated by 1996, along with carbon tetrachloride and methyl chloroform, while the phaseout of halons was to be achieved by 1994. Among the more contentious issues, the phaseout date for HCFCs was brought forward to 2020, although essential use exemptions were included in the agreement, reflecting strong lobbying mainly from the U.S. chemical firms and refrigeration and air conditioning sector. The U.S. delegation argued successfully that HCFC use in air conditioners for large buildings be permitted until 2030, a position that received the support of France. To the dismay of environmentalists, the world's leading HCFC producers, DuPont and Elf Atochem, had succeeded in securing a largely industry-friendly outcome on the question of transitional substances.

It became clear after the Copenhagen conference that the major regulated CFCs had reached the end of their lifeline. Industry in the developed countries had made sufficient progress for the CFC phaseout deadline to be brought forward even further. The EC environment ministers decided only weeks after Copenhagen to set the year 1995 as the new CFC phaseout date (*IER* 1993a). This was followed by EPA's announcement of a proposed regulation that would ban most CFC uses by the end of the same year (*IER* 1993b). The CFC producers had long stopped opposing shorter phaseout deadlines and concentrated now on securing a sufficient lifetime for their substitute chemicals. The announcement in June 1993 of an earlier phaseout for HCFCs in Europe, therefore, caused some concern on the part of European producers. But the Commission's proposal to achieve total HCFC elimination in 2014, rather than 2030 as agreed in Copenhagen, applied only to domestic consumption in the EC and did not stop European producers of HCFC to continue exporting the chemicals to other countries that still relied on the transitional substitute (*IER* 1993c).

The firms that were most threatened by this development were not the CFC producers, but the user industries. While the major CFC producers and policymakers in Europe and North America cooperated in seeking to eliminate CFC production at the earliest opportunity, some of the user industries that had delayed conversion efforts now faced a situation of rapidly dwindling CFC supplies. Their new problem was that America's CFC producers were planning to stop producing CFCs much earlier than anticipated—in the case of DuPont in 1994.

It was the refrigeration and air conditioning sector that was hit hardest by the speed of the CFC phaseout. In the United States, car manufacturers reacted to the ensuing crisis by lobbying the government to grant exemptions under the "essential use" rules that would allow continued CFC usage. The American Automobile Manufacturers Association, a powerful grouping of America's car makers, pointed out that 140 million cars were in use that needed future supplies of CFCs to service their existing air conditioning systems. U.S. car makers were only planning to start introducing new systems with HFC-134a as a coolant in 1994, the very year that DuPont planned to stop manufacturing CFCs. As a consequence, the automobile industry faced a serious squeeze on CFC stocks in the near future, which in turn would result in higher servicing costs for millions of car owners (*IER* 1993d).

Given the political sensitivity of the issue, the U.S. government gave in to industry pressure and approached DuPont with a request to extend CFC production by one year. The move proved to be highly embarrassing for the government and the chemical producers, as both had so far cooperated in speeding up the CFC phaseout. In the end, lobbying by the car industry and fears of a voter backlash against higher servicing costs won the day (*IER* 1994a).

By the time the Seventh Meeting of the Parties was convened in Vienna in December 1995, the focus in international negotiations had shifted from the CFC phase-out program to debates on international aid to developing countries (Falkner 1998) and the inclusion of previously unregulated ODS, particularly the thorny issue of methyl bromide. On the question of the HCFC phaseout, industry's position had received a boost in the latest Technical and Economic Assessment Panel report published before the Vienna Conference. The panel concluded that although technically feasible, an earlier phaseout date of 2015 would cause unjustifiably high economic costs, as the HCFC-using refrigeration and air conditioning equipment would have a lifetime far beyond 2015 (*IER* 1995). The revisions of the HCFC regulations agreed in Vienna reflected the prevailing expert opinion. Although the phaseout date was officially moved from 2030 to 2020, a "service tail" of ten years was included that allows industrialized countries to supply existing equipment with HCFCs. The decision was strongly criticised by environmental groups for reflecting industry needs rather than environmental concerns (Greenpeace 1996, 4). Yet again, a coalition of HCFC producer and user interests prevailed in the negotiations.

Conclusion: Corporate Power in Ozone Politics

There can be little doubt that leading chemical firms played a key role in bringing about a rapid elimination of CFC emissions, once an international agreement to protect the ozone layer had been reached. They did this not out of altruism but in response to societal and political pressures. But this does not mean that the business response was merely reactive or epiphenomenal. It assumed a more direct political significance as the parties to the Montreal Protocol sought to strengthen the regulatory regime and make its implementation work. Based on their pivotal role in directing technological change, corporations were able to influence the design and phasing of the protocol's regulatory instruments.

To be sure, corporations were not "in control" of the treaty revision process; an economic reductionist account would not adequately reflect the reality of international environmental politics, in ozone layer protection or elsewhere. The corporate sector did not "dictate" the terms and conditions for regulating ODS. The ozone regime was very much a second best solution, particularly for the CFC producers, who had fought over ten years against regulation and incurred high costs in switching to CFC substitutes. But technological power gave corporations the edge over other actors in shaping the regulatory discourse that unfolded as the implementation of the ozone regime got underway.

The political role of corporations in the Montreal Protocol process has at least two dimensions. At a fundamental level, business support for the ozone regime helped to legitimise the role of the CFC producers in the search for technological alternatives to ozone depleting chemicals. Once an international agreement was reached in 1987—and leading industry players themselves had a role to play in this—the CFC industry came to be seen not simply as part of the problem, but also as part of the solution. Governments listened to business advice and actively sought to engage corporate actors in the international political process. Some industry experts even became part of the influential UNEP technology assessment panels that gave authoritative advice to governments on the technical hurdles to phasing out ODS.

Furthermore, as the parties to the Montreal Protocol moved from norm setting to implementation, the corporate sector's technological power became a critical factor in international ozone politics. It allowed the corporate sector to shape the evolution of the ozone regime and exert influence over governmental actors well beyond their lobbying clout. Technological power, understood as the power to direct technological innovation and its introduction in the market, had a direct impact on the regulatory discourse. It set the parameters for what actors perceived as technologically feasible regulations, and thus framed the knowledge structure within which the ozone regime evolved.

Corporations did not control, in the strict sense of the word, the process of technological change. Technological innovation is not, however, an exogenous phenomenon governed by the random accumulation of scientific knowledge. Decisions on investment in alternative CFC technologies, combined with the market power that CFC producers or users possess, can give rise to a considerable degree of influence over the direction of technological change. This was evident in the case of the chemical industry that had decided to invest in transitional substances, such as HCFCs, which it sought to protect against the demands for an early phase-out under the Montreal Protocol's revised regulations.

It is important to recognize that corporate power in the case of ozone politics often found its match not only in the agency of other actors (states, NGOs) but also through business competition and conflict. The CFC producers faced several challenges in their attempt to secure a viable market for their substitute products, as their control over user industries varied from sector to sector. In some cases, predominantly refrigeration and air conditioning, the chemical industry was able to build on its close relationship with CFC users and secure a sizeable market for its preferred substitutes HCFC and HFC. In contrast, the aerosol manufacturers very quickly abandoned any attempt at introducing transitional substances and switched to hydrocarbon technology, thus making the chemical industry's preferred option redundant. Competition, and lack of coordination among business actors, prevented a uniform political front of CFC producers and users. As so often in the relationship between the private and public sector, business conflict creates space for other political actors (Falkner 2001).

In addressing the problem of ozone layer depletion, the CFC industry therefore never found a win-win solution possible. It could only hope to delay regulatory action, as it did during the 1970s and early 1980s, or to shape the emerging regulatory regime to make it more business-friendly, as became the predominant business approach from the late 1980s onwards. In any case, the chemical industry lost a profitable niche market and the user industries had to invest in costly product and manufacturing process redesigns. The case of business involvement in ozone politics clearly does not lend itself to political-economic explanations that are rooted in economic determinism. In fact, a variety of factors account for the creation, and success, of the Montreal Protocol: political leadership by states and individual state leaders; scientists' efforts to promote understanding of ozone layer depletion and its link to CFC emissions; environmentalist campaigns that mobilized public opinion; and international organisations that provided a forum for negotiation and implementation. But as this chapter argues, corporate leadership in the negotiations and search for substitute technologies came to play a critical role in shaping the process with which the broad coalition of actors in favour of international regulations began to phase out ozone-depleting substances.

Notes

1. The Montreal Protocol did, of course, create an uneven incentive structure for phasing out ODS in that it did not regulate all ozone-depleting chemicals from the outset. By covering only five CFCs (11, 12, 113, 114, 115) and three Halons (1211, 1301, 2402)—a compromise reflecting the delicate balance between precautionary action and commercial interests—the negotiators of the Montreal Protocol created a regulatory framework that allowed users of unregulated substances (e.g., HCFCs, methyl chloroform, carbon tetrachloride) to delay action for many years. But within the group of regulated substances, the protocol did not differentiate between different usage types.

2. DuPont's CFC business, although providing a reliable source of profit, was not central to the company's overall business strategy. Profit margins in CFC production were below average, particularly since the onset of the ozone controversy and the collapse of the CFC aerosol business in the United States. CFC sales accounted for only 2 percent of DuPont's total revenues in 1987 (Reinhardt 1989, 10).

3. Even DuPont had only declared a goal of complete CFC phase out by 2000 (*WSJ* 1989).

References

ACHRN (1989, April 24). Government understands industry's needs on CFCs. *Air Conditioning, Heating & Refrigeration News*, p. 28.

Alliance for Responsible CFC Policy (1989). *HCFCs and HFCs Provide the Balance*. Washington, D.C.: Alliance for Responsible CFC Policy.

Ayres, E., & French, H. (1996). The refrigerator revolution. *World Watch*, September/October, 14–21.

Benedick, R. E. (1991). *Ozone diplomacy: New directions in safeguarding the planet*. Cambridge, MA: Harvard University Press.

Chemical Engineering (1988a, January 18). Chemical firms search for ozone-saving compounds. *Chemical Engineering*, pp. 22–25.

Chemical Engineering (1988b, April 25). Firms ally to speed development of CFCs. *Chemical Engineering*, pp. 37–38.

Cook, E. (1996). *Marking a milestone in ozone protection: Learning from the CFC phase-out*. Washington, DC: World Resources Institute.

Der Spiegel. (1988, October 24). Es geht darum, unsere Haut zu retten. *Der Spiegel*, 42(43), 40–46.

Der Stern (1989, May 18). Dicke Gewinne mit Ozonkillern. *Der Stern*.

European Chemical News (1987, September 28). Atochem criticizes terms of global CFC agreement. *European Chemical News*, p. 5.

European Chemical News (1989, February 6). ICS-Kali in CFC link, DuPont patents blends. *European Chemical News*, p. 17.

ENDS (1987, September). Firms push substitutes for ozone-depleting aerosols. *ENDS Report, 152*, 4.

ENDS (1989, January). Majority of electronics firms lack plans to replace CFCs. *ENDS Report, 168*, 5–6.

EPA (1997, August). Champions of the world: Stratospheric ozone protection awards. EPA430-R-97-023. Washington, DC: Environmental Protection Agency.

Falkner, R. (1998). The multilateral ozone fund of the Montreal Protocol. *Global Environmental Change, 8*(2), 171–175.

Falkner, R. (2001). Business conflict and U.S. international environmental policy: Ozone, climate, and biodiversity. In P.G. Harris (Ed.), *The environment, international relations, and U.S. foreign policy* (pp. 157–177). Washington, DC: Georgetown University Press.

FAZ (1988, September 26). Umweltfreundlicherer Bau von Kühlschränken. *Frankfurter Allgemeine Zeitung*, p. 16.

FAZ (1989, July 6). Ein Schiff voller Demonstranten. *Frankfurter Allgemeine Zeitung*, p. 6.

Glas, J. P. (1988). DuPont's position on CFCs. *Forum for Applied Research and Public Policy, 3*(3), 71–72.

Greenpeace (1996). Radically accelerated phase out of methyl bromide and HCFCs with controls on HFCs: An environmental imperative. Greenpeace International Position Paper, prepared for the Eighth Meeting of Parties to the Montreal Protocol. November, Costa Rica.

Haas, P. M. (1992). Banning chlorofluorocarbons: Epistemic community efforts to protect stratospheric ozone. *International Organization, 46*(1), 187–224.

IER (1988, February 10). Companies should plan on bigger cuts in CFC production, consumption, NRDC says. *International Environment Reporter*, p. 111.

IER (1989, March). EC Council agrees on total ban of CFCs by 2000; Bush says U.S. goal dovetails. *International Environment Reporter*, p. 105.

IER (1992, November 18). Montreal Protocol nations expected to speed phase-out of ozone depleters. *International Environment Reporter*, pp. 731–732.

IER (1993a, January 13). Ministers approve stepped up timetable to phase out ozone depleting substances. *International Environment Reporter*, p. 7.

IER (1993b, March 24). EPA proposes listing methyl bromide as ozone depleter slated for phase out. *International Environment Reporter*, p. 209.

IER (1993c, June 16). EEC proposal sets earlier phase-out date for HCFCs than Montreal Protocol requires. *International Environment Reporter*, p. 428.

IER (1993d, June 30). Automakers seek options to accommodate phase-out of coolants that deplete ozone. *International Environment Reporter*, p. 500.

IER (1993e, July 14). Propane/Butane, HFCs, other chemicals being tried by German refrigerator makers. *International Environment Reporter*, pp. 525–526.

IER (1994a, January 12). Environmentalists criticize request by United States for more CFC production. *International Environment Reporter*, pp. 22–23.

IER (1994b, May 4). Hoechst becomes first chemical company to stop production of chlorofluorocarbons. *International Environment Reporter*, p. 390.

IER (1995, November 29). Montreal Protocol parties should weigh risks, not just costs, Greenpeace says. *International Environment Reporter*, pp. 903–904.

Jachtenfuchs, M. (1990). The European Community and the protection of the ozone layer. *Journal of Common Market Studies, 28*(3), 261–277.

Jordan, A. (1997). The ozone endgame: The implementation of the Montreal Protocol in the UK. *CSERGE Working Paper*, GEC 97-16. Norwich.

Kauffman, J. M. (1997). Domestic and international linkages in global environmental politics: A case-study of the Montreal Protocol. In M. A. Schreurs & E. C. Economy (Eds.), *The internationalization of environmental protection* (pp. 74–96). Cambridge, UK: Cambridge University Press.

Levy, D. L. (1997). Business and international environmental treaties: Ozone depletion and climate change. *California Management Review, 39*(3), 54–71.

Litfin, K. (1994). *Ozone discourses. Science and politics in global environmental cooperation.* New York: Columbia University Press.

Makhijani, A., & Gurney, K. R. (1995). *Mending the ozone hole: Science, technology, and policy.* Cambridge, MA: MIT Press.

Manufacturing Chemist (1988, September). Industry gears up to find CFC alternatives. *Manufacturing Chemist, 59*(9), 9.

Maxwell, J. H., & Weiner, S. L. (1993). Green consciousness or dollar diplomacy? The British response to the threat of ozone depletion. *International Environmental Affairs, 5*(1), 19–41.

Oye, K. A., & Maxwell, J. H. (1995). Self-interest and environmental management. In R. O. Keohane & E. Ostrom (Eds.), *Local commons and global interdependence: Heterogeneity and cooperation in two domains* (pp. 191–221). Newbury Park, CA: Sage.

Pollack, A. (1991, May 15). Moving fast to protect ozone layer. *New York Times,* p. D1.

Reinhardt, F. (1989). DuPont Freon products division. Washington, D.C.: National Wildlife Federation.

Roan, S. (1989). *Ozone crisis: The 15-year evolution of a sudden global emergency.* New York: John Wiley & Sons.

Somheil, T. (1996, October). Time to optimize. *Appliance,* pp. 29–33.

SZ (1988, December 7). Hoechst will Ozonschicht schützen. *Süddeutsche Zeitung.*

United Nations Environmental Programme (UNEP) (1989a, June 30). *Aerosols, sterilants and miscellaneous uses of CFCs.* Report, Technical Options Committee on Aerosols, Sterilants and Miscellaneous Uses.

United Nations Environmental Programme (UNEP) (1989b, November 13). *Synthesis report.* UNEP/OzL.Pro.WG.II(1)/4.

United Nations Environmental Programme (UNEP) (1989c, June 30). *Technical progress on protecting the ozone layer.* Report of the Technology Review Panel.

WSJ (1989, October 26). Hoechst AG wants to have CFC replacement by 1995. *The Wall Street Journal,* p. 4.

Zurer, P. S. (1988, February 8). Search intensifies for alternatives to ozone-depleting halocarbons. *Chemical & Engineering News,* p. 18.

6

The Genetic Engineering Revolution in Agriculture and Food: Strategies of the "Biotech Bloc"

Peter Andrée

Introduction

Do international relations precede or follow... fundamental social relations? There can be no doubt that they follow... (Gramsci 1971, 176)

In the mid-1990s, when transgenic seeds[1] were first being planted commercially in North America, the biotechnology industry assumed that these crops would become the food of the future, providing a growing population with improved nutrition and farmers with more sustainable production options (Duvick 1995). Given these ambitions, the agricultural biotechnology revolution appears to be a mixed success to date. On the one hand, between 1995 and 2002 the area planted in transgenic seeds grew from a few test plots to 58.7 million hectares (James 2002), an enormous achievement for the champions of biotech seeds. However, 99 percent of this coverage is in only four countries: the United States (66 percent), Argentina (23 percent), Canada (6 percent), and China (4 percent). Elsewhere, an effective moratorium exists on the planting of new genetically modified organisms (GMOs) in member states of the European Union, and several countries have considered banning GMOs altogether (including Sri Lanka, Croatia, and Bolivia) (Villar 2002). A number of African states even rejected food aid because it contains GM grains (Vint 2002). Furthermore, Japan, Australia, and the members of the EU, among others, have created laws that will require the labelling and/or "traceability" of GMOs through the food system. At the international level, the Cartagena Protocol on Biosafety—a treaty under the auspices of the Convention on Biological Diversity designed to protect the environment and human health from risks that may be caused by international trade in (living) GMOs—entered into force in September 2003. It allows countries to deny shipments of GMO seeds and commodities on precautionary grounds, even if their risks are not fully substantiated by scientific evidence. These are all signs that the global acceptance of GM crops and foods is far from assured.

The uncertain status of the biotech revolution deserves closer examination. What forces are driving—and resisting—the biotech revolution in agriculture? And, what is the role of multilateral governance structures in this field? To address these questions I adopt a neo-Gramscian framework. In keeping with the introductory quote from Gramsci, this approach involves an examination of the "fundamental social relations" that have enabled the introduction of GMOs in the global agriculture and food system, and which will ultimately determine their fate.

Since the 1970s, the techniques of genetic engineering offered the promise of new possibilities for agriculture, but a neo-Gramscian approach would suggest that a global biotech "revolution" requires much more than new technologies. This revolution is here characterized as part of a protracted "war of position," a revolution led by the biotech industry but in concert with a multitude of alliances and supportive shifts within both the economy and society. While the biotech coalition had achieved a stable, secure position, or hegemony, in the United States and several other states by the late-1990s, it remains contested internationally.

In order to analyze the construction of hegemony, Gramsci argues that we must look for supportive shifts within the "relations of force" at three levels: the level of material forces of production; the relations of political forces through which the interests of one class fraction come to be accepted as the common interest of society in general; and the relations of military forces and other coercive actions on the part of governments (Gramsci 1971, 181–184). The bulk of this chapter systematically examines these three layers of (interrelated) strategies adopted by the biotech industry and its allies in order to bring GMOs into global agriculture.

When the balance of force at all three levels of relations converge around a specific and coherent set of ideas, we have in place what Gramsci would call an "historical bloc." As Levy and Newell point out, however, Gramsci and his contemporary interlocutors use this term in two different ways. It is sometimes used to refer to the alliances among social forces necessary to move a particular agenda forward. Elsewhere in Gramsci's writings, "historical bloc" describes the alignment of material, organizational, and discursive formations that stabilize relations of production and meaning and thus enable these alliances. This is the understanding of historical bloc that informs Gill's work on the "disciplinary neoliberalism" associated with the rise of the "transnational historical bloc" since the early 1970s (Gill 1998). Levy and Egan (2003), in their study of the responses of U.S. oil and automobile industries to international climate change negotiations, demonstrate the relationship between the two concepts. They analyze how shifts in the material, organizational, and discursive spheres undermined the hegemony of the fossil fuel coalition in the United States.

I draw on the first reading of historical bloc in my definition of the "biotech bloc," which is spearheaded by a handful of agrichemical corporations, but which also involves promotional and regulatory arms of government and civil society institutions (such as universities), that have worked together to realize a particular vision of genetic engineering in agriculture. I draw on the second reading to explain the impact of wider (already hegemonic) institutions and ideologies on the development of the biotech bloc. In this case, my focus is on the implications of the transnational historical bloc and its ideology of neoliberalism on the way that agbiotech has developed over the last three decades.

The theoretical approach of this chapter is very different from the stance taken in most writings on "regimes" in the global environmental politics literature. This work demonstrates, through the example of biotechnology, that regulatory structures represent only one dimension of governance. The scientific practices of genetic engineering, corporate product strategies, and the responses of farmers and consumers to GMOs, are equally important in defining the way that societies govern nature in this field. Still, formal regulatory structures are important and need to be examined in the context of larger patterns of governance. Regulatory practices that place controls on industrial practices are often established in order to build consent among groups in society who are suspicious of the direction taken by the leaders of a hegemonic formation. In some cases, these mechanisms represent only minor accommodations towards critics, as we see in the GMO regulatory structures established in the United States and Canada. At other times, the shape taken by regulatory structures may signal larger concessions given to the critics of an emergent hegemonic formation. The Cartagena Protocol on Biosafety appears to be an example of the latter.

Situating the Biotech Bloc within Neoliberalism

The first successful genetic engineering of a strain of bacteria took place in California in 1972. The first incorporation of foreign genes into the DNA of plant cells, and then the propagation of plants from them, occurred in the early 1980s. Over the period of this decade, the context of genetic engineering research underwent significant transformations.

Initially, this branch of molecular biology was pursued by academic scientists, funded by government research councils—primarily in the United States and in Britain (Wright 1994). In the mid 1970s, as researchers became aware of the potential economic benefits of their new engineering science, many began biotech start-up companies to research products and techniques. From 1975 to 1983, more than 250 small biotechnology firms were founded in the United States, bankrolled by venture capital (Shand 2001, 223–224). This number was much higher than in any other country or region of the world, establishing the United States as the center for research into commercial applications of biotechnology.

In the late 1970s and early 1980s, large pharmaceutical and agrichemical corporations such as DuPont, Pfizer, and Monsanto began to move into the field, starting in-house biotech R&D projects as well as undertaking partnerships with, or simply buying out, biotech start-ups. These agrichemical companies were lured by the potential of genetic engineering to provide new products and to reshape existing markets. By 1981 in the United States, it was said to be difficult to find a recombinant DNA practitioner who did not have an industrial connection (Wright 1994, 104).

This brief sketch shows how the biotech bloc coalesced long before the first commercial products became available, involving a potent combination of science, capital, government, and corporate support. While this bloc had global dimensions (including participation from European firms), it was strongest in the United States (along with Argentina and Canada), because this is where a close alignment was first achieved across the three levels of relations of force, as described later. The development of agricultural applications of biotechnology in the 1980s occurred in the context of the ascendancy of a global regime of accumulation, centered around transnational capital and the ideology of neoliberalism. The structures of production, ideologies, and institutional forms of this larger historical bloc help explain the successful establishment of the biotech bloc in the Americas.

At an ideological level, the acolytes of neoliberalism argue that the only true role for government is the establishment and enforcement of the rule of law, the main purpose of which is to assign and protect the rights of individuals to property and economic freedom (von Hayek 1976, 101–104). The growing strength of this perspective was fundamentally important to the biotech bloc's emergence through a set of decisions made in U.S. courtrooms and in the U.S. patent office. In 1980, the U.S. Supreme Court ruled that a transgenic microorganism could be assigned a patent in the *Diamond v. Chakrabarty* decision. This patent ruling was followed in 1983 by a similar U.S. Patent and Trademark Office judgement regarding transgenic plants, and another in 1987 involving a transgenic mouse (Shand 2001). These rulings gave the emerging biotech industry a major incentive to continue their explorations. Evidence for this can be found in the way that U.S. biotech start-up stocks soared in the fall of 1980 following the initial patent decision (Charles 2001, 11).

These rulings also represented a significant philosophical shift. In his majority decision in the Chakrabarty case, Chief Justice Burger wrote that any matter that has been subject to useful improvement should fall under the 1793 Patent Act, which is based on the idea that "ingenuity should receive a liberal encouragement" (Burger 1980). However, plants and animals were not included in the original patent law (King and Stabinsky 1998/1999). Mr. Justice Brennan, who wrote the dissenting opinion in the case, took note of this, and the fact that over time the United States had developed legislation, including the 1970 and 1978

Plant Variety Protection (PVP) Acts, which gave more limited protection over biological innovations (Brennan 1980; Ghijsen 1998). In these PVP Acts, the rights of a company to exclusively market seeds of a plant line it had developed are balanced with the "farmer's privilege" to replant seed, as well as the rights of other breeders to develop varieties from the protected line. The 1980 Chakrabarty ruling demonstrated a swing towards private and corporate interests in interpretations of intellectual property law in the United States. This was premised on the idea that the technological ability to reengineer an organism is somehow special, requiring a higher degree of protection than the laborious practices of plant breeding. While seed companies had sought for years to extinguish the farmer's privilege, it took the mystique of genetic engineering to achieve their goal.

When it came to opposition to patenting of living organisms, the Chakrabarty case did raise the ire of early U.S. biotech activists such as Jeremy Rifkin. However, the larger U.S. environmental organizations did not mount public campaigns against it. As the case involved a strain of bacteria engineered to help clean up oil spills, it was seen as an environmentally friendly technological development by these groups (Mooney 1996). While there is no evidence that the Chakrabarty case was specifically chosen for its public relations potential, the example does show that the perceived benefits of a particular genetic transformation can help mitigate public dissent for the biotech project as a whole.

Neoliberal arguments concerning the desirability of "free" markets are suspect, given that markets are embedded in legal and institutional structures that are themselves the product of ideological currents and political compromises; this is demonstrated by the way that property rights over new plant varieties have changed over time. Still, an entrenchment of the "imperative" to follow "independent" market signals enabled important shifts in research structures and norms, which defined the direction taken by the biotech bloc in the United States and Canada. First, this argument allowed industry funding of academic research to grow. The harbinger of this trend was a groundbreaking 1975 Monsanto-Harvard Medical school agreement, in which Harvard received $23 million over ten years. In return, Monsanto received first rights to all patentable results of the project. By 1982, the emerging biotech industry invested

$326 million per year in U.S. universities and investment rates were rising at an annual rate of 11 percent (Wright 1994, 58). Second, government research sponsorship patterns changed. North American government research councils began actively promoting industry/university partnerships in the 1980s. In Canada, this would result in a "matching funds" model, whereby public funds were prioritized for researchers who could find matching funds from industry (Agriculture and Agri-Food Canada 1996).

These structural changes to the funding of university research had several impacts. While public and corporate money would be targeted at developing the engineering capabilities of molecular biology, little research on the ecological and other impacts of GMOs was undertaken (Crouch 1991). The fact that biotech research was market driven also affected the choice of products to be developed. By the mid-1980s, researchers learned that although they could use these techniques to develop, for example, virus-resistance in certain plants, the agenda would be driven by the search for traits that could bring major economic returns (Charles 2001, 31–40).

Discourses of national competitiveness have also had a considerable impact on the development of biotechnology, especially in the context of debates over regulation. While critics raised questions about security, ethics, health, and safety, these concerns were juxtaposed against a driving need for "competitiveness" (Gottweis 1998). In the United States, in particular, arguments that "overly cautious" regulation could result in "delays in achieving benefits" were taken very seriously by the Reagan administration (Wright 1994, 275). Even representatives from Monsanto and Calgene (the two companies to first develop commercial biotech products in the United States), who wanted government regulation in order to convince the public of the safety of GMOs, faced an uphill battle (Eichenwald et al. 2001). Key administrators within the U.S. Food and Drug Administration argued that the larger biotech companies only wanted regulation to exclude from the market smaller competitors, who could not afford the expense of getting a product over the required regulatory hurdles (Miller 1997). When the case for regulation was finally heard, the resultant framework operated through two key mechanisms: First, a voluntary system of notifications and standards was put in place for GMOs rather than mandatory premarket authorizations (Food and Drug Administration 1992; Office of Science and Technology Policy 1986); second, this system was calibrated to address risks that are "real and significant rather than hypothetical or remote" (OSTP 1992, 6761–2). The result was a system that assumed the essential "familiarity" and "substantial equivalence" of most GMOs to non-GMOs. On the one hand, this regulatory minimalism allowed the United States to achieve its goal of becoming the base for the global agricultural biotech industry. Even European companies such as Hoechst would come to set up shop on American shores. On the other hand, this approach further reduced the incentive for scientists to pursue research into the ecological and health implications of GMOs (Meyer 1998).

As the transnational historical bloc grew in influence over the last three decades, it had to face challenges from a variety of quarters, and it has had to make some concessions. However, Gramsci is careful to point out that "such sacrifices and such compromise cannot touch the essential . . . in the decisive nucleus of economic activity" (Gramsci 1971, 212). The environmentalist challenge to the transnational historical bloc is an important case in point. As contributions to this volume demonstrate, transnational corporations (TNCs) have made some accommodations to environmentalist critiques, but they have adopted a range of strategies to minimize the impact upon underlying material interests. One common strategy is to argue that a company's activities should be seen as part of the solution to the goal of "sustainable development" rather than the problem. Arguing that certain genetically modified crops could result in less use of dangerous herbicides and pesticides, while providing greater yields, leaders in the biotech sector have worked hard to frame their activities within the discourse of sustainable development (Shapiro 1999).

Despite such arguments, however, the biotech bloc has had difficulty convincing environment and development civil society organizations (CSOs), as well as many officials from less developed countries, of the fit between biotechnology and sustainable development. The record of the Green Revolution, which enhanced short-term food yields at the cost of significant social and environmental problems, led many to believe that biotechnology would follow a similar path. Furthermore, the "bio-prospecting" by genetic engineering firms for genetic material was seen by many in the South as "biopiracy," a theft of resources from the world's poorest countries (Shiva 1997).

The contested relationships between biotechnology, the environment, and global equity emerged at the inter-governmental level in negotiations leading up to the Convention on Biological Diversity (CBD) in Rio in 1992. While many biotechnology and agrichemical corporations, as well as the United States, supported a vision of "sustainable development" at Rio, their reaction to the CBD demonstrated their underlying position on these issues. The administration of George H. W. Bush refused to sign the CBD because of its clauses on the "fair and equitable" sharing of the benefits arising from biotechnologies based upon resources from Southern countries (Munson 1993, 503).

Each of these examples shows how the biotech bloc was shaped in the context of the ascendancy of the transnational historical bloc and its neoliberal interpretations of property rights, the role of markets, regulation, and sustainable development. As the biotech bloc grew in strength and ambition, some of its specific interests would also come to shape the activities of the larger historical bloc. For example, central to the liberal project are international efforts to institutionalize Lockean-style property rights through the WTO and its agreement on Trade Related Intellectual Property Rights (TRIPS; Gill 1998). Among its provisions, the TRIPS requires countries to create systems for protecting the developers' intellectual property rights (IPR) for new microorganisms and plants developed through genetic engineering, through patents or an effective *sui generis* system. This demand for patents clearly serves the interests of TNCs, including agrichemical companies. In fact, executives from these companies admit to drafting this agreement in the early 1990s (under the General Agreement on Tarriffs and Trade (GATT; Shiva 1998). However, in order to have the TRIPs accepted, these companies did have to make concessions in the final text. The mention of *sui generis*, meaning "of its own kind of class," was included to appease developing countries who were not comfortable with the idea of patenting living organisms. How this is to be interpreted lies at the heart of ongoing debates over IPRs in biotechnology.

Material Foundations of the Biotech Bloc

The following three sections look more closely at the biotech bloc itself, presenting an overview of the material, discursive, and organizational strategies employed by this bloc in the particular field of the agrifood system. While I deal with these sets of strategies sequentially, it is important to recognize that they are all closely intertwined, as the examples demonstrate.

One of the first commercial applications of genetic engineering of crops—the one which continues to account for 83 percent of all commercial GM seeds planted (James 2002)—was the introduction of genes that allow plants to be resistant to broad-spectrum herbicides. This application, more than any other, shows how the commercial biotech project is driven by the interests of agrichemical companies, for it allowed them to extend the market reach of their chemical product lines.

When agrichemical companies first started producing transgenic seeds, most did not actually own seed companies. In order to get the herbicide-tolerance trait to farmers, new relationships were developed by these companies with both seed companies and farmers. Monsanto was the first to establish a new form of licensing to seed companies. When Monsanto licensed the right to use the herbicide-resistance genes to seed companies, the agreement stated that each bag of seed would be printed with the phrase "Roundup Ready." This was a technique that the company had successfully employed in marketing its aspartame sweetener "Nutrasweet" to soft-drink companies. With Nutrasweet, once consumers came to identify with its logo on a can of cola, Monsanto was able to raise the price it exacted from the drink manufacturers. However, despite the company's hopes to develop the market for Roundup-Ready crops in the same way, initial deals between Monsanto and Pioneer Hi-Bred show that the Roundup Ready gene in soybeans gave little return to Monsanto aside from its use in selling more chemicals (Charles 2001, 120–121).

The second most important application of genetic engineering to date (found in one-quarter of globally produced transgenic crops (James 2002), are corn, potato, and cotton seeds engineered with genes from the bacteria *Bacillus thuringiensis* (Bt) to express proteins that are toxic to a variety of insect pests. These seeds allowed the biotech purveyors to take market share from their agrichemical competitors, because they are designed to replace the need for insecticides. They also present potential environmental problems, however. Bacterial forms of Bt have historically been used as "natural" pesticides, and scientists worried that widespread production of Bt would result in the rapid development of insect resistance to this valuable pesticide. Biotech companies rejected this argument, but were eventually forced, first by the U.S. Department of Agriculture (USDA) and then by similar government regulatory actions in Canada and elsewhere, to accept a "refuge" strategy that would see farmers plant up to 20 percent of their fields with nonbiotech crops to slow the development of insect-resistance (National Research Council 2000). Thus, the advent of Bt crops provides an example of the way the biotech industry was forced to make accommodations in order to win government and civil society consent.

In marketing Bt crops, Monsanto and others granted licenses in exchange for a specific premium to be added to the price of the seed. This "Technology Fee" was designed to be a little less than the costs saved by the farmer by using the gene (such as costs of insecticides or more expensive herbicides) (Charles 2001, 122). This arrangement allowed the biotech companies to keep direct control of the price charged to farmers for their genes, as well as the right to enforce their property rights. When buying biotech seeds, farmers would then sign a Technology Use Agreement (TUA) directly with the agrichemical company. These agreements actually amounted to a new form of contract farming. Since the 1970s, a contract farming model—in which food processing companies supply all of the necessary inputs, and offer a guaranteed price for the output—had been gradually taking over independent farming in a variety of agricultural sectors in the United States and Canada (notably poultry and potatoes) (Friedmann 1994), but no inroads had been made into the dominant field crops of soybeans and corn. With genetically modified seeds, the contract model arrived for these field crops, although without a guaranteed return on the farmers' investment. The TUAs specify that farmers will only use that company's herbicides on the crop (even if generic equivalents are available), that they will not save any seed for replanting, and that the biotech company has the right to inspect their farms and grain bins up to three years after planting (Monsanto 2000).

The TUAs led to considerable resentment towards biotechnology from many in the farming community. Farmers recognized the new contracts as a power-grab on the part of the biotech purveyors, which reduced the competition for farm inputs and thus raised their prices. Still, the contracts did not stop many farmers from buying into GE seeds. This was largely because herbicide-resistance and pesticidal crops were generally effective, and considered time-savers for farmers. The crops also brought economic benefits to the first wave of farmers to adopt them. One study found that the relative returns for switching to Bt cotton in 1997 were almost as high for the farmers as they were for the company (Falck-Zepeda et al. 1999). Unfortunately, these increased returns for GMOs are unlikely to remain the norm. The structural consequences of adopting GMOs can already be seen in Argentina, where farmers began growing herbicide-tolerant soy in the mid 1990s; commodity prices declined while reliance on imports of chemicals went up. The overall number of producers decreased, and average farm size increased (Lehmann and Pengue 2000).

TUAs, and their enforcement through North American courts, have worked to ensure that farmers buy new seed each year. Believing that such legal means of protecting intellectual property may not be as robust in some developing country markets, a U.S. biotech company (Delta and Pine Land) developed a series of genetic modifications designed to shut off the reproductive ability of a plant, so that the seeds from a transgenic plant could not reproduce successfully. This was termed the Technology Protection System (TPS), because it "simplifies protection of technology and removes it from the legal arena" (Radin 1999).

Although the U.S. patent for TPS was awarded in 1998, the trait has not yet been introduced in a commercial seed variety, partly because this proposed modification has created more of a public backlash than any other trait. In India, as one example, over 80 percent of farmers currently plant their own farm-saved seeds (Swaminathan 1998). When the patent on the trait was made public in 1998 by the Canadian-based Rural Advancement Foundation International (RAFI), environmental, human rights, and development groups around the world criticized this "terminator technology," holding it up as proof that the emerging biotechnology industry had set its sights on removing the rights of the world's farmers to save and reseed their crops. The outcry was so great that late in 1999 the CEO of Monsanto—which earlier that year had made an offer to buy Delta and Pine Land—was forced to state publicly that his company would not commercialize the TPS. This appears to be another example of how the biotech industry was forced to accommodate public pressure. However, in March 2000, insiders in Delta and Pine Land's labs told RAFI that they have never slowed their development efforts for terminator seeds. RAFI has also found that over thirty other patents have been issued in the United States to thirteen different institutions describing techniques that control seed germination (Rural Action Foundation International 1999).

Biotechnology has been so successful in the Americas that it has utterly changed the global seed business. While in the early 1990s, agrichemical companies were licensing their biotech traits, by the end of the decade most had undertaken outright buyouts of seed purveyors. For example, Monsanto bought out U.S. seed giant Asgrow in 1996 followed by Plant Breeding International, the leading seed company in Great Britain, in 1998. Meanwhile, DuPont bought out the largest seed company in the United States, Pioneer Hi-Bred, in 1999 (Charles 2001, 195, 235). By the autumn of 1999, five companies (AstraZeneca, DuPont, Monsanto, Novartis, and Aventis) accounted for the entire global transgenic seed market, two-thirds of the global pesticide market, and almost one-quarter of the global seed market (Shand 2001, 231).

Despite these major consolidations in the global farm input industry, it is important to note that there are even larger titans in the food system, and these may, ultimately, have the final say about the future of food biotechnology. These are the giant food companies and grain shippers. The largest food company in the world, Nestlé, had 1997 revenues of $45.3 billion, larger than the entire commercial seed industry ($23 billion) and the entire agrichemical industry ($31 billion) (Shand 2001, 229). Even grain trading giants such as Archer Daniels Midland (ADM) and Cargill dwarf Monsanto.

To date, the reaction from the food giants and grain traders has been mixed. On the one hand, many in the food industry look forward to new traits in food products that enhance their appeal to consumers. To this end, they have worked with the biotech companies to get minimalist labelling standards for GM foods and other supportive government policies. In line with this, Heffernan (1999) argues that we are beginning to see the global consolidation of the agricultural input industry with food processors and distributors into three to four major "food system clusters" that will control food from the gene to the dinner plate. Such clusters would likely continue to develop common strategies to support biotech.

On the other hand, when European consumers—followed by some of their North American counterparts—widely rejected GM foods in the late 1990s, a number of food processors, including McCain's in Canada as well as food retailers such as Marks and Spencer in Britain, decided to remove GM ingredients from their products. ADM sent out a letter warning grain terminals that they may have to segregate GM soy (Charles 2001, 164). A similar scenario played out in early 1999, when Europeans were hesitant about accepting GM corn (Charles 2001, 254). In the first case, Europeans did decide to accept GM soy, so costly and complicated segregation was not necessary. In the second case, American corn was diverted elsewhere, so segregation was again not needed. Still, these examples show that the interests of the food processors and grain traders, who are more sensitive to consumer demand, are not necessarily the same as those of the agricultural input industry. This continues to be a dangerous scenario for the seed purveyors.

Discursive Strategies of the Biotech Bloc

Gramsci notes that every relationship of hegemony is necessarily an educational relationship (Gramsci 1971, 350). He terms "organic intellectuals" those individuals and groups, emerging from within an historical bloc, who are able to establish a justificatory framework for change. In the case of the biotech bloc, the organic intellectuals were the scientists, agricultural college administrators, and government regulators who were the product of the relationships between academia, industry, and government of the 1970s and 1980s. They included Donald Duvick, for a time the head of research at Pioneer Hi-Bred, who wrote articles justifying the biotech revolution in academic journals using his Iowa State University address (see Duvick 1995). When resistance to biotechnology was peaking in 1999, twenty-six deans of the major U.S. agricultural colleges signed a National Agricultural Biotechnology Council (NABC) "Vision for Agricultural Research and Development in the twenty-first Century" proposal that supported the biotech industry's agenda, including, among other things, application of the TPS/terminator technology (NABC 2000, p. 189).

In order to be successful, Gramsci argues that organic intellectuals must be able "to transcend a particular form of common sense and to create another which [is] closer in conception of the world of the leading group" (Gramsci 1971, 423). We can identify four central discursive strategies employed by the biotech bloc to establish its project as the new common sense. Most pervasive are attempts to situate genetic engineering within the powerful discourses of progress, modernization, and later, "sustainable development." In one industry-sponsored booklet, food biotechnology (the industry typically avoids the term genetic engineering) is defined as "diverse activities from the use of yeast in brewing or bread-making to advanced plant-breeding techniques" with a history that dates back to 1800 BCE (Council on Biotechnology Information 2000). In this way, the techniques of genetic engineering are framed as part of a set of beneficial innovations with a long and uncontroversial history.

A second common discursive strategy emphasizes the precision and predictability of "modern" biotechnology, which is claimed to involve the "precise" identification and transfer of "the specific gene that creates a desired trait in a plant" to another plant (CBI 2000). This portrayal is designed to maximize public trust by minimizing the uncertainties and complexities involved in the science. It relies on a reductionist interpretation of DNA as the genetic "code" or "blueprint of life." Such metaphors suggest that each strand of DNA translates directly into a single phenotypic expression, regardless of the organism or environment in which it is found. This account of genetics was crucial in developing the early U.S. regulatory framework for GM foods, which asserts that GMOs can be "generally recognized as safe" because, in most cases, they simply involve the addition of a specific, known trait to a familiar food plant (FDA 1992). It was also successfully mobilized to argue against any distinct labelling of GMO products in Canada and the United States.

Keller terms this discourse of molecular genetics "gene talk" (Keller 1995). While many of the experiments undertaken by molecular biologists do show that a strand of DNA has a predictable function when inserted into another organism, Keller stresses that gene talk is only a partial reading of the evidence provided by biological research. Genes can interact in complex, poorly understood ways, so the same "gene" sometimes does very different things in different organisms (Commoner 2002; Holdrege 1996).

In regulatory circles, the gulf between the "precision" and the "uncertainty" framings of genetic engineering have crystallized in a debate around the call for "sound science" in GMO regulation. On the one side, proponents of biotech (including North American regulators) argue that, given the higher degree of precision in transgenic techniques, and the fact that there are as yet no known cases of unexpected allergens or toxins produced by GMOs, they should only be kept off the market when there is clear scientific evidence that they are harmful; GMOs should be presumed safe unless proven otherwise. On the other side, antibiotech forces have converged around calls for much more "precautionary" biotech regulation, suggesting a shift in the burden of proof: Because of the unpredictable nature of genetic engineering, GMOs should be proven safe before they are widely adopted. Some advocates for precaution call for an outright ban on GMOs (e.g., Myhr and Traavik 1999), while others use the principle to call for more stringent regulatory systems that carefully evaluate all of the anticipated and unanticipated effects of the engineering process on GMOs prior to their commercial release (Royal Society of Canada 2001). In Europe, a more precautionary reading of the risks of GMOs influenced the development of the 1990 EC Regulatory Directive (90/220) on the Deliberate Release of GMOs to the environment, and an even stronger interpretation of precaution shaped the 2001 update of this directive (2001/18/EC) (European Parliament 2001). Because of the growing influence of the precautionary principle on European law, the debate between sound science and precaution has become a key point of contention among the United States, Argentina, and Canada, on the one hand, and the EU on the other.

At the multilateral level, criticism regarding the presumed safety of GMOs (as embodied in the U.S. regulatory framework) was one factor that led developing countries to insist on the need for a legally binding international agreement reaffirming their right to restrict imports of GMOs for the sake of protecting biodiversity and human health in their countries, and that would help them build capacity to regulate GMOs. These demands were first raised during the negotiations of the 1992 UN Convention on Biological Diversity (Munson 1993). A clause calling for consideration of the need for such a protocol was successfully placed in the CBD in 1992. In 1996, negotiations on a "Biosafety Protocol" to the CBD began. The biotech bloc's response to the challenges posed by these negotiations is discussed below.

Where the first three discursive strategies employed by the biotech bloc emphasize a "biotechnology continuum" and the higher precision and safety of its latest techniques, the final set of discursive strategies stresses the uniqueness of genetic engineering; if genetic engineering is to realize its benefits, it is argued that GMOs must be given a higher level of proprietary protection.

The contradiction between this emphasis on uniqueness and other claims for GMOs' similarity to previous techniques has not been lost on critics. In fact, the idea of higher levels of proprietary protection over genetically modified organisms (and even simply genomic "translations" of DNA) has become a major point of resistance to the biotech bloc. Northern CSOs such as Genetic Resources Action International (based in Spain) and RAFI (now ETC, based in Canada) have made the call for "no patents on life" the primary focus of their activities, mounting an international campaign against the TRIPS agreement in particular. Noting that the TRIPS allows *sui generis* systems other than patenting, more moderate positions argue for the protection of the farmers privilege (often termed "farmer's rights") to save and use seed varieties, as well as the rights of indigenous communities to compensation for stewardship of plant varieties and their genetic material (Louwaars 1998).

Various international fora have become sites of this struggle, and critics have made some gains. For example, the 2001 International Treaty on Plant Genetic Resources for Food and Agriculture ensures that some traditional crop varieties will be excluded from patentability (Food and Agricultural Organization 2001). We are also seeing *sui generis* laws enacted in developing countries, which recognize the rights of traditional plant breeders and farmers to register plant varieties used and developed by them or their communities as part of IPR legislation. This is the intention, for example, of India's Plant Varieties and Farmers' Rights Bill, passed in 2001 (Center for Science and Environment 2001). Notably, none of these initiatives negates the ability of biotechnology corporations to claim IPRs over "invented" transgenic plant varieties.

Organizational Strategies for Internationalizing the Biotech Revolution

At the international level, the biotech bloc has employed an array of organizational strategies to build and maintain support in the face of criticism. Given the centrality of government participation for biotech hegemony in the United States, the bloc first tried to get the backing of other governments. And since developing countries often fall into line once industrialized countries achieve a consensus on technical issues, the focus was initially on building common regulatory positions among developed countries.

Beginning in the mid-1980s, consensus-building discussions took place through the Organisation for Economic Cooperation and Development (OECD), an organization with a history of supporting trade liberalization by promoting regulatory harmonization. The OECD brought together biotech experts and government representatives to try to develop a common perspective on how to assess the risks of GMOs (Organisation for Economic Cooperation and Development 1986, 1993). Although a "consensus" was usually achieved in these talks, the results were actually highly ambiguous regulatory definitions for terms such as "substantial equivalence," which could then be interpreted very differently in diverse regulatory cultures (Andrée 2002). Still, through the 1990s the biotech bloc was able to use OECD processes to entrench internationally ideas and terms originating in the minimalist U.S. regulatory system.

A variety of other transnational organizational strategies were targeted at building a European environment hospitable to the introduction of

GMOs. These included the activities of a working group of the Trans-Atlantic Business Dialogue—an organization of European and North American businesses that informally develops trade policy proposals for streamlining regulatory hurdles in the United States and Europe (Trans-Atlantic Business Dialogue 2002)—which put forward a proposal to set up a European counterpart to the U.S. FDA to accelerate and harmonize the approval of new products (Levy and Newell 2000: 18). An EU–U.S. biotechnology group was also set up under the umbrella of the Transatlantic Economic Partnership (TEP), a more formal process established in 1998 to work towards a binding trans-Atlantic trade and investment agreement (United States Trade Representative 1998).

Such actions to encourage regulatory harmonization were supplemented with coercive pressures on European countries to import GMOs. For example, when U.S. farmers started growing Roundup-Ready soy in the spring of 1996, and these were not yet approved for import into Europe, the United States sent Special Trade Representative Mickey Kantor (later a Monsanto board member) to argue that if Europe denied imports of soy, the United States would consider an appeal under the WTO (Charles 2001, 186).

Initially, these efforts to get GMOs into Europe were successful. Even though Greenpeace tried to block shipments of GM soy in the fall of 1996, the EU bowed to U.S. pressure and allowed the imports. This was due, in part, to the fairly strong relationships the biotech bloc had built with several European governments, including Britain, Germany, and France (Steffenhagen 2001). However, by June 1999 public resistance against GMOs was so strong that Environment Ministers from France, Italy, Greece, Denmark, and Luxembourg declared that, pending the adoption of stricter regulations, they would suspend further authorizations for the growing or marketing of GMOs (European Council and Commission 1999).

Given that the biotech bloc had included European companies from the outset, why had its fortunes turned on that continent? We can find clues at each of Gramsci's three levels. At the level of production, it is noteworthy that even in 1999 European seed companies were not as committed to the techniques of genetic engineering as their American counterparts. A survey of ninety-nine European seed companies showed that they had only earmarked 38 percent of their R&D budgets to develop crops using these techniques (Arundel et al. 2000). At the organizational level, European Civil Society Organizations (CSOs) critical of GMOs were very good at mobilizing public skepticism about agricultural biotechnology. One key tool for them was the 1997 decision by the European Parliament to require food containing GMOs to be labelled as such. Furthermore, the regulatory process in the EU allowed more room for political intervention by national governments in approvals than would be the case in the United States or Canada. Finally, at the discursive level, the arguments for biotechnology's advantages over traditional agriculture were generally not convincing in Europe. In France and Germany, for example, there is a strong attachment to a countryside inhabited by small farms and wildlife. In this context, GMOs were easily perceived as an American development designed to make small farmers beholden to TNCs introducing potential new risks to the environment. Such risks were made highly tangible in the spring of 1999 (just before the European moratorium was announced) when a Cornell researcher published a study in *Nature* showing that Bt corn varieties, already approved in the United States and in Europe, could be harmful to endangered monarch butterfly populations in the United States (Losey et al. 1999).

While each of these factors helps to explain the European rejection of GMOs, the crux of the difference between North America and Europe relates to the bovine spongiform encephalopathy (BSE, or "Mad Cow" disease) scandal in Britain in the spring of 1996—the very year that Roundup-Ready soy was set to arrive on the continent—and the ability of CSOs to draw connections between the unknowns associated with Mad Cow case and "frankenfoods." On March 20th, after years of official denials of the possibility of such an occurrence, British Prime Minister John Major announced that at least ten people had died from a human form of that disease and that the victims likely acquired the disease from eating British beef. As Charles notes, "from that moment the burden of proof, when it came to the safety of Europe's food, had firmly settled on the shoulders of the biotechnology industry" (Charles 2001, 170).

In the wake of heightened resistance to its products, the biotech bloc has used international negotiations concerning the safety of GMOs to try to maximize gains in friendly fora, such as the Codex Alimentarius and the OECD, and to minimize losses in more hostile settings, such as the Biosafety Protocol negotiations. The Codex, originally formed under the UN Food and Agricultural Organization (FAO) and the World Health Organization (WHO), has become accepted, through the WTO, as the global standard setting body for food safety. In March of 2002, lengthy debates within a Task Force on issues related to GMOs resulted in a statement of principles that provide a global framework for evaluating the safety and nutritional aspects of GMOs. It also provides for guidelines on labelling of GMOs (Codex 2002). At first glance, the biotech bloc appears to have won in these meetings, because the Task Force upheld many of the common regulatory practices of GMO exporters, such as the policy of only labelling GMOs if they can be shown to present a credible health risk. But this should come as no surprise. Among international organizations, the Codex is known to be strongly influenced by transnational food and agrichemical companies. Between 1989 and 1991, for example, Nestlé had thirty-eight representatives on national delegations to Codex committees. Next best represented were Coca-Cola, Unilever, and then Monsanto (Avery et al. 1993).

However, to state that the biotech bloc "won" here is to oversimplify. The outcomes of these meetings actually called for higher regulatory standards than are currently in place in the United States. For example, these principles call for greater transparency than is currently practiced there, where much of the data upon which regulatory decisions are based is considered "confidential business information." These principles also state that, for the foreseeable future, no GMO can be considered "conventional," thereby escaping regulatory scrutiny, a practice already accepted in the United States (Strauss 2002). A similar pattern can be seen within the OECD. Since the strong European backlash to biotech, the OECD's activities appear to have become focused on developing a compromise between more precautionary European regulatory approaches and the U.S. approach. This is evident in a 2000 OECD report, which examines how the concept of substantial equivalence can actually be employed in a scientifically robust way (OECD 2000).

These examples suggest that the biotech bloc is being forced to make accommodations—even in fora historically favorable to neoliberal values—that within the context of the United States and Canada alone they were able to avoid. In fora that emerged in the context of criticism of biotechnology, such as the negotiations for the Biosafety Protocol, the bloc appears to have been forced to make even larger concessions.

By the time the Biosafety Protocol was being negotiated in the mid-1990s, some national governments had already made a strong commitment to biotechnology—namely the United States, Canada, and Argentina, or were on the verge of doing so (Australia). In clear examples of the bloc's successful hegemony at the domestic level, these countries saw their national interests as being tied to the future export of GMOs and wanted minimal interference to such exports. Wanting to show that a trade-oriented position did not only come from "GMO exporters," these countries worked to build allies for their position, courting Uruguay and Chile in particular. The establishment of the alliances that would become the *Miami Group* in the Protocol negotiations is an instructive example of the organizational efforts of the biotech bloc within developing countries. Canadian biotech experts, including nutritionists initially trained in the art of selling biotech's safety by the Monsanto-sponsored *Dietician's Network* (Stewart 2002), were sent to Chile and Uruguay to run information sessions for their government representatives (BIOTECanada 1999). Costs for these trips were covered by Canadian government agencies, until they realized they could be criticized for supporting such blatant efforts to promote biotechnology (Abley 2000).

Aside from working through certain states, the biotech bloc also acted in the Biosafety Protocol negotiations through industry organizations, which participated as formal observers. These groups occasionally made interventions in the talks and were constantly engaged in lobbying in the hallways, backrooms, and cocktail circuits. The strategy here was to develop coalitions among biotechnology companies around the world in an attempt to show, once again, that biotechnology is not simply in the interests of the United States and Canada. Thus was born the Global Industry Coalition (GIC), which was said to represent more than 2,200 firms from over 130 countries worldwide (GIC 1999). Notably, the GIC had no formal presence outside the Biosafety Protocol negotiations. It was based in Canada and represented by spokespersons for the Canadian biotechnology lobby group BioteCanada, an organization that received over Can$6 million from the Canadian government between 1994 and 1999 (Abley 2000).

During the Biosafety negotiations, the Miami Group, along with GIC and other industry organizations, fought for the least restrictive measures possible. Initially, they argued for purely voluntary guidelines—as established through the United Nations Environment Program (UNEP) in 1995 (UNEP 1995)—but this debate was lost when European governments decided to support the call of developing countries for a legally binding instrument. When faced with the prospect of a binding instrument, the biotech bloc argued, once again, for "sound science" as the only acceptable basis of any restriction on the import of a GMO. They also wanted minimal interference in the global trade in GMO commodities destined for food, feed, or processing, and they demanded a "saving's clause" that would subordinate the Protocol to relevant WTO treaties.

Sensing growing international resistance to GMOs, in November 1999 the United States and Canada tried to adopt the strategy of "forum shifting" (Braithwaite and Drahos 2000, 28), by taking their case directly to the WTO talks in Seattle. Knowing that controls would likely be stronger under the Biosafety Protocol, the final negotiations of which were about to take place in Montreal, fourteen Environment Ministers from European countries flew to Seattle to publicly contradict the position of their trade commissioner by rejecting the need for new talks under the WTO, and to reiterate their support for the Biosafety Protocol process. This effectively killed the North American proposal (Winfield 2000).

At the final Biosafety negotiation session, the Miami Group was forced to concede on a number of key points in the final hours of negotiation, rather than be seen to be unwilling to accept global consensus on GMO regulation. A number of contextual factors, including the *Nature* report on Bt corn and Monarch butterflies, the public rejection of GMOs in Europe and the spread of protests to North America and Japan, strong civil society representation at the talks, and media reports condemning the Canadian position (often orchestrated through civil society groups), were all influential in producing this result.

On the surface, it appears that the Cartagena Protocol (the formal name given to the final text) was a victory for groups critical of GMOs. It does guarantee the rights of nation states to restrict imports if they believe these organisms could affect biodiversity within their country. Although negotiations are not yet complete, it may also lead to a liability regime that deters biotech purveyors and strict shipping documentation requirements that could lead to segregated commodity streams for GMOs.

However, if we dig deeper, we see that the biotech bloc did not fair poorly in the final result. Many classes of GMOs are not covered by the protocol, including anything not "living," all "products" of GMOs (such as corn oil), and pharmaceuticals. Bulk commodity shipments have a separate, simplified procedure for import approvals. Furthermore, the interpretation of "precautionary" decision making in practice is still vigorously debated by scientists and policy makers. The World Trade Organization (WTO), for example, has been reticent to accept certain readings of precaution as the basis for rejecting imports (see for example, the beef hormone ruling against the EU [WTO 1998]), so much depends on the relationship between the WTO and the Protocol. Unfortunately, the Protocol's language on this relationship does not clarify matters. The preamble to the final compromise text states that: (1) trade and environment agreements should be mutually supportive; (2) that the Protocol should not be interpreted as implying a change in the rights and duties of a Party under other international agreements; but (3) that this does not mean that the protocol is subordinate to other agreements.

In the area of biosafety, coercion is, as a Gramscian framework would suggest, also a mechanism through which the bloc is countering strict biotech regulations around the world. Consider the cases of countries who have proposed bans on GMOs: In 2001, Croatia announced its intention to ban the import and growing of GMOs as part of its drive to be the ecotourist destination of Europe. In response, the United States threatened to take Croatia to the WTO. Similar threats led the Sri Lankan government to indefinitely defer the draft 2001 Food Act, which would ban GMOs (Villar 2002). Most recently, in May 2003, the United States launched a WTO case against the EU, arguing that its GMO moratorium is not based on "sound science" and that it is harmful to those developing countries who have followed the EU's lead in rejecting GMOs (United States Department of Agriculture 2003).

Conclusions

The neo-Gramscian framework employed here brings to light the diversity of material, discursive, and organizational strategies that can be adopted by a group seeking to have its interests accepted as the common interest. Domestic and international regulatory structures, often simplistically characterized as "compromises among stakeholders," crystallize in the context of the "war of position" waged by a leading group trying to establish a hegemonic formation. The extent to which these structures represent real compromises is only evident through detailed analysis of the economic and political stakes involved.

In the United States and Canada, the close relationships realized in the biotech sector offer a strong example of the power of hegemony at the domestic level. While institutional and discursive strategies helped to establish this, the choice of the first generation of genetically engineered products was also clearly important. Even though herbicide-resistance and Bt-genes contributed to skepticism about the benefits of genetic engineering, they did allow for a return on investment. At a time when there was still uncertainty concerning the profitability of GMOs, these traits brought returns by changing the rules of the game in markets that the early biotech purveyors knew well: the herbicide and insecticide markets. These were also traits that would appeal to farmers. Given the controversial strategies developed to retain control of the engineered genes (the TUAs), this was critical. Finally, if the first biotech crops had been resistant to more toxic chemicals, such as the herbicide Atrazine, sold by Ciba-Geigy, the environmental politics of GMOs may have played out very differently in the United States and Canada.

The effectiveness of working with governments to win consent for GMOs in North America led the biotech bloc to adopt similar strategies in Europe. However, in Europe the bloc underestimated the ability of public-interest CSOs to mobilize public skepticism against GMOs. The resultant European rejection of GMOs has put the biotech bloc in a difficult position. The United States has initiated WTO actions against the EU for its moratorium on GMOs, but this action could cause the bloc more harm than good for two reasons. First, this challenge will place both regulatory systems under international scrutiny. Given the U.S. system's reliance on simplistic assumptions about the equivalence between GMOs and non-GMOs, and given that scientific criticism of these assumptions has grown through the 1990s, it is not certain that the U.S. regulatory approach could weather international scrutiny. Second, even if the United States wins its case, European states may not bow to a WTO panel ruling that goes against such strong public sentiment. In the beef hormone case, even after the EU lost, they did not open up markets to beef produced with hormones from North America (accepting economic sanctions instead) because the dispute itself had simply added fuel to the fire of public rejection.

This is not meant to suggest that the global biotechnology revolution is likely to fail. Rather, regulatory struggles appear to have forced a shift in the biotech bloc's expectations. Through the Cartagena Protocol, OECD harmonization processes, the Codex Alimentarius and the more precautionary domestic regulations being codified in Europe and else-where, the bloc has had to make concessions in return for clarity on the global governance of GMOs. The bloc will use openings provided to it (e.g., capacity building processes under the Protocol), but firms may also be held to a higher burden of proof regarding the safety of GMOs, and this will certainly slow the pace of the biotech revolution. With the Pro-tocol in force, the institutional structures are also in place to justify major restrictions on the trade of specific GMOs should evidence emerge demonstrating significant harms associated with them. The possibility of such restrictions (and the consequent need for segregation) mean that GMOs may yet become a liability for the grain traders, just as evidence of harm in GM foods could result in these becoming a liability for the food processing and retailing titans. Were either case to arise, reactions from these players may yet cripple the biotech bloc's emerging, but fragile, global hegemony.

Note

1. Although there are technical differences, I use the terms biotechnology, genetic engineering, genetic modification, and transgenesis interchangeably to refer to the techniques designed to modify the genetic material of an organism in intentional ways, often through the incorporation of DNA from other species.

References

Abley, M. (2000). Biotech lobby got millions from Ottawa: Public cash used to alter image. *The Montreal Gazette*, p. A1.

Agriculture and Agri-Food Canada (AAFC) (1996). *Agriculture and agri-food: Canada's action plan*. Ottawa: Government of Canada.

Andrée, P. (2002). The biopolitics of genetically modified organisms in Canada. *Journal of Canadian Studies*, 37(3), 162–191.

Arundel, A., Hocke, M., & Tait, J. (2000). How important is genetic engineering to European seed firms? *Nature Biotechnology*, 18, 578.

Avery, N., Drake, M., & Lang, T. (1993). *Cracking the Codex: An analysis of who sets world food standards*. London: National Food Alliance.

Beachy, R. N. (1991). The very structure of scientific research does not mitigate against developing products to help the environment, the poor, and the hungry. *Journal of Agricultural and Environmental Ethics*, 4(1), 159–165.

BIOTECanada (1999). *Public Awareness and Risk Assessment in Agricultural Biotechnology*, Santiago: AgBiotechNet. <www.agbiotechnet.com/proceedings/May%202000/chilpub.asp>.

Braithwaite, P., & Drahos, J. (2000). *Global business regulation*. Cambridge, UK: Cambridge University Press.

Brennan, W. (1980). Dissenting opinion in the case of *Diamond v. Chakrabarty*, 447 U.S. 303. Available at <http://www.law.uh.edu/healthlaw/law/Federal Materials/FederalCases/Diamondv.Chakrabarty.htm> (accessed March 12, 2004).

Burger, W. (1980). Opinion of the Court in the case of *Diamond v. Chakrabarty* 447 U.S. 303. Available at <http://www.law.uh.edu/healthlaw/law/Federal Materials/FederalCases/Diamondv.Chakrabarty.htm> (accessed March 12, 2004).

Centre for Science and Environment (CSE) (2001). Seeds of discontent? India's first legislation aimed at protecting farmers' rights raises many apprehensions. *Down to Earth* (New Delhi), 10, 6, 48–51.

Charles, D. (2001). *Lords of the harvest*. Cambridge, MA: Perseus Publishing.

Codex (2002). Report of the Codex Ad Hoc intergovernmental task force on foods derived from biotechnology—Third session. Yokohama, Japan: Codex Alimentarius Joint FAO/WHO Food Standards Programme.

Commoner, B. (2002). Unraveling the DNA myth: The spurious foundation of genetic engineering. *Harper's Magazine*, February, 1–15.

Convention on Biological Diversity (CBD). (2000). Cartagena protocol on biosafety to the convention on Biological Diversity. Montreal: Secretariat of the Convention on Biological Diversity.

Council on Biotechnology Information (CBI) (2000). *Biotechnology: Good ideas are growing*. Ottawa, Ontario: Council for Biotechnology Information.

Crouch, M. (1991). The very structure of scientific research mitigates against developing products to help the environment, the poor, and the hungry. *Journal of Agricultural and Environmental Ethics*, 4(2), 151–158.

Duvick, D. (1995). Biotechnology is compatible with sustainable agriculture. *Journal of Agriculture and Environmental Ethics*, 8(8), 112–125.

Eichenwald, K., Kolata, G., & Peterson, M. (2001, January 25). Biotechnology food: From the lab to a debacle. *New York Times*, p. A1.

European Council and Commission (1999). Minutes of the Council Meeting 24–25 June. Annex: Declaration regarding the proposal to amend directive 90/220/EEC on Genetically Modified Organisms. Brussels.

European Parliament (2001). Directive 2001/18/EC on the deliberate release into the environment of genetically modified organisms and repealing Council Directive 90/220/EEC. *Official Journal of the European Communities*, L(106), 0001–0038.

Falck-Zepeda, J. B., Traxler, B., & Nelson, R. G. (1999). Rent creation and distribution from the first three years of planting Bt cotton, Vol. 14. New York: The International Service for the Acquisition of Agri-biotech Applications (ISAAA) Briefs. Available at <http://www.isaaa.org/Publications/briefs/briefs_14.htm> (accessed March 12, 2004).

Food and Agricultural Organisation (FAO) (2001). International Convention on Plant Genetic Resources for Food and Agriculture Approved by FAO Conference. Rome: United Nations FAO (press release 01/81 C5).

Food and Drug Administration (FDA) (1992). Statement of policy: Foods derived from new plant varieties. *Federal Register*, 57. Available at <http://vm.cfsan.fda.gov/~lrd/biocon.html> (accessed March 12, 2004).

Friedmann, H. (1994). Distance and durability: Shaky foundations of the world food system. In P. McMichael (Ed.), *The global restructuring of Agro-Food systems* (pp. 258–276). Ithaca, NY: Cornell University Press.

Ghijsen, H. (1998). Plant variety protection in a developing and demanding world. *Biotechnology and Development Monitor*, 36, 2–5.

Gill, S. (1998). New constitutionalism, democratisation, and global political economy. *Pacifica Review*, 10(1), 23–40.

Global Industry Coalition (GIC) (1999). Biodiversity Jeopardized in Cartagena Biosafety Negotiations. Cartagena, Columbia: Global Industry Coalition (press release). Available at <http://www.iatp.org/iatp/News/news.cfm?News_ID=142> (accessed March 12, 2004).

Global Industry Coalition (GIC) (2001). Capacity Building Efforts of Individual Biotechnology Companies. Biosafety Clearing House, Convention on Biological Diversity. <www.bch.biodiv.org/Pilot>.

Gottweis, H. (1998). *Governing molecules: The discursive politics of genetic engineering in Europe and the United States.* Cambridge, MA: MIT Press.

Gramsci, A. (1971). *Selections from the prison notebooks.* New York: International Publishers.

Heffernan, W. (1999). Biotechnology and mature capitalism. In D. P. Weeks, J. B. Seglken, & R. W. F. Hardy (Eds.), *World food security and sustainability: The impacts of biotechnology and industrial consolidation.* Ithaca, NY: National Agricultural Biotechology Council. Available at <http://www.cals.cornell.edu/extension/nabc/pubs/pubs_reports.html#nabc11> (accessed March 12, 2004).

Holdrege, C. (1996). *A question of genes: Understanding life in context.* Hudson, NY: Lindisfarne Press and Floris Books.

James, C. (2002). Global status of commercialized transgenic crops: 2002. Ithaca, NY: International Service for Agricultural Acquisitions of Agri-Biotech Applications. Available at </www.isaaa.org>.

Keller, E. F. (1995). *Refiguring life: Metaphors of twentieth century biology.* New York: Columbia University Press.

King, J., & Stabinsky, D. (1998–1999). Biotechnology under globalisation: The corporate expropriation of plant, animal and microbial species. *Race and Class,* 40(2–3), 73–89.

Lehmann, V. (1998). Patent on seed sterility threatens seed saving. *Biotechnology and Development Monitor, 35,* 6–8.

Lehmann, V., & Pengue, W. A. (2000). Herbicide tolerant soybean: Just another step in a technology treadmill? *Biotechnology and Development Monitor, 11*(43), 11–14.

Levy, D. L., & Egan, D. (2003). A neo-Gramscian approach to corporate political strategy: Conflict and accommodation in the climate change negotiations. *Journal of Management Studies, 40*(4), 803–830.

Levy, D. L., & Newell, P. (2000). Oceans apart? Business responses to global environmental issues in Europe and North America. *Environment, 42*(9), 9–20.

Losey, J. E., Rayno, L. S., & Carter, M. E. (1999). Transgenic pollen harms Monarch larvae. *Nature 399,* 6733.

Louwaars, N. P. (1998). Sui generis rights: From opposing to complementary approaches. *Biotechnology and Development Monitor, 36,* 13–16.

Meyer, H. (1998). Precise precaution versus sloppy science: A case study. Presentation given at the 5th meeting of the Open-Ended Ad Hoc Working Group on Biosafety, Montreal (on file with author).

Miller, H. (1997). *Policy controversy in biotechnology: An insider's view.* Austin, TX: R. G. Landes Co.

Monsanto (2000). *Technology use agreement terms and conditions: Roundup-ready canola.* Monsanto Canada Inc. (on file with author).

Mooney, P. R. (1996). Civil and uncivil societies. *The commodification of life.* A. K. Welch. Vancouver, B.C.: The Pomelo Project, Simon Fraser University.

Munson, A. (1993). Genetically manipulated organisms: International policy-making and implications. *International Affairs, 69*(3), 497–517.

Myhr, A. I., & Traavik, T. (1999). The precautionary principle applied to deliberate release of genetically modified organisms (GMOs). *Microbial Ecology in Health and Disease, 11,* 65–74.

National Agricultural Biotechnology Council (NABC) (2000). *The bio-based economy of the twenty-first century: Agriculture expanding into health, energy, chemicals and materials.* Ithaca, NY: Author.

National Research Council (NRC) (2000). Genetically-modified pest-protected plants: Science and regulation. Washington, DC: National Academy Press.

Office of Science and Technology Policy (OSTP) (1986). Coordinated framework for regulation of biotechnology. *Federal Register, 51.* Available at <http://usbiotechreg.nbii.gov> (accessed March 12, 2004).

Office of Science and Technology Policy (OSTP) (1992). Exercise of federal oversight within the scope of statutory authority: Planned introductions of biotechnology products into the environment. *Federal Register, 57,* 22984–2 3005.

Organisation for Economic Cooperation and Development (OECD) (1986). *Recombinant DNA safety considerations.* Geneva: OECD.

Organisation for Economic Cooperation and Development (OECD) (1993). Safety evaluation of foods derived by modern biotechnology: Concepts and principles. Paris: OECD.

Organisation for Economic Cooperation and Development (OECD) (2000). Report of the Task Force for the Safety of Novel Foods and Feeds. Geneva: OECD.

Purdue, D. (1995). Hegemonic trips: World trade, intellectual property and biodiversity. *Environmental Politics, 4*(1), 88–107.

Radin, J. (1999). The Technology Protection System: Revolutionary or evolutionary? *Biotechnology and Development Monitor, 37,* 24.

Rissler, J., & Mellon, M. (1996). *The ecological risks of engineered crops.* Cambridge, MA: MIT Press.

Royal Society of Canada. (2001). *Elements of precaution: Recommendations for the regulation of food biotechnology in Canada.* Ottawa: Royal Society of Canada.

Rural Action Foundation International (RAFI) (1999). Traitor tech: The terminator's wider implications. Ottawa: RAFI <www.rafi.org>.

Shand, H. (2001). Gene giants: Understanding the "life Industry." In B. Tokar, *Redesigning Life?* (pp. 222–237). London: Zed Books.

Shapiro, R. (1999). How genetic engineering will save our planet. *The Futurist, 33*(4), 28–29.

Shiva, V. (1997). Bioethics: A Third World issue. New Delhi: Research Institute for Science, Technology and Ecology. <http://www.nativeweb.org/pages/legal/shiva.html>.

Shiva, V. (1998). Monocultures, monopolies and the masculanisation of agriculture. New Delhi: Research Foundation for Science, Technology and Ecology. <http://www.nativeweb.org/pages/legal/shiva.html>.

Schmidheiny, S. (1992). *Changing course.* Cambridge, MA: MIT Press.

Steffenhagen, B. (2001). *The influence of biotech industry on German and European negotiation positions regarding the 2000 Cartagena protocol on biosafety.* Unpublished thesis. Otto-Suhr-Institut fur Politikwissenschaft. Berlin: Freie Universitat Berlin.

Stewart, L. (2002). Good PR is growing. *This Magazine,* May/June, 5–10.

Strauss, S. (2002, March 26). Truce called in "frankenfood" fight. *Globe and Mail* (Toronto), R6.

Swaminathan, M. S. (1998). Farmers' rights and plant genetic resources. *Biotechnology and Development Monitor, 36,* 6–9.

Trans-Atlantic Business Dialogue (TABD) (2002). Homepage <www.tabd.org>.

United Nations Environment Program (UNEP) (1995). International Technical Guidelines for Safety in Biotechnology. Nairobi: UNEP.

United States Department of Agriculture (USDA) (2003). U.S. and Cooperating Countries File WTO Case Against EU Moratorium on Biotech Foods and Crops. USDA Press Release 0156.03. <http://www.usda.gov/news/releases/2003/05/0156.htm>

United States Trade Representative (USTR) (1998). United States and European Union Conclude Joint Action Plan for the Transatlantic Economic Partnership. Washington, D.C.: Office of the United States Trade Representative. <http://www.ustr.gov/regions/eu-med/westeur/98-99.pdf>.

Villar, J. L. (2002). U.S. steps up pressure to force acceptance of genetically modified organisms worldwide. Friends of the Earth International Briefing Paper, released at the Third Meetings of the Intergovernmental Committee for the Cartagena Protocol (ICCP-3). April 22–26, The Hague, The Netherlands.

Vint, R. (2002). Force-feeding the world. UKabc. <www.UKabc.org/forcefeeding.htm>.

von Hayek, F. A. (1976). *The road to serfdom.* London: Routledge.

Wright, S. (1994). *Molecular politics: Developing American and British regulatory policy for genetic engineering, 1972–1982.* Chicago: University of Chicago Press.

World Trade Organisation (WTO) (1998). EC measures concerning meat and meat products (hormones): Report of the Appellate Body (AB-1997-4). Geneva: WTO.

III

Business Influence: Regional Dimensions

7

The Environmental Challenge to Loggers in the Asia-Pacific: Corporate Practices in Informal Regimes of Governance

Peter Dauvergne

Just fifty years ago old-growth forests blanketed much of the tropical Asia-Pacific. Since then corporate loggers have irrevocably degraded most of the valuable commercial forests. The Philippines, for example, once one of the world's largest tropical timber exporters, is now a net importer of tropical timber, with old-growth forests covering less than three percent of total land area. The outcry over this ecological devastation has put loggers under increasing pressure over the last decade to harvest on a genuinely sustainable basis. Internationally, a forests regime has been consolidating with the principle of sustainable management at the core. Nationally, states have been reforming policies to increase timber revenues and internalize environmental principles. Such changes would appear at first glance to pose a great challenge to corporate loggers in the Asia-Pacific. Yet, so far, the transformation of the rules and principles of forest management has done little to alter logging practices on the ground.

Why this disjuncture? This chapter argues that strategic bargains and compromises between loggers and their sociopolitical allies are diluting and deflecting pressures for genuine environmental reforms. Loggers maintain informal networks of private alliances with state officials (politicians, bureaucrats, and military officers), societal power brokers, and other corporate executives (firms along the trade chain, including affiliates and financiers). The aggregate effect of such networks is to steer environmental reforms toward the formal rules of governance—international agreements, public corporate language and, most commonly, state policies—and away from the informal elements—the expected and accepted norms that arise from webs of state-business alliances. There is no doubt that some of these reforms have strengthened formal rules. It is true, too, that some reforms to formal rules have hurt the financial interests of loggers (especially stricter national policies to increase a government's share of timber revenues). Yet, overall, the process of reform is primarily widening the disjuncture between the rules/rhetoric (including corporate) and actual norms/practices on the ground. In some countries, such as Indonesia, forest management is even getting worse, with the rates of deforestation and illegal logging on the rise. There is no corporate conspiracy here, nor even elaborate cooperation among firms. Rather, the widening disjuncture is an outcome of the collective impact of the reactions of individual firms within a sociopolitical setting where networks of allies benefit personally and professionally from the current arrangements.

Such a process of change is delaying rather than completely blocking reforms. Firms nevertheless perceive this as a strategic victory. The reason is straightforward: firms see future access to valuable commercial timber, in terms of both volumes and markets, as highly uncertain. A firm's primary goal is therefore to maximize immediate profits. There are many sources of uncertainty, both real and imagined, from fluctuations in tropical timber prices to conflicts between locals and loggers. The nature of the informal logging rules are themselves often the greatest source of uncertainty. It is common, for example, for a firm to lose access to a site after the fall of a political, military, or local patron. It is common, too, for new political leaders to label the timber allies of old leaders as "corrupt," sometimes even jailing them. With such uncertainty facing firms, and given that loggers will deplete most of the valuable commercial forests in twenty years or so anyway, a delay to far-reaching reforms is as good as a "win." Before proceeding to defend these arguments with a closer analysis of the global system of forest governance as well as the local rules and practices in the Asia-Pacific, the chapter begins by explaining the theoretical significance of this case study for understanding the role of business in global environmental governance.

Corporations and Global Environmental Governance

This book is an example of the growing academic interest in trying to better understand the role of corporations in global environmental governance. Recent studies point to important changes to corporate procedures and rhetoric, sometimes for ethical reasons, sometimes for financial reasons, and sometimes in response to NGO or consumer campaigns (*IDS Bulletin* 1999; Paterson 2001; Rowlands 2000). One common theme in the literature is that firms are working hard to weaken environmental regulations or subvert environmental campaigns (Rowell 1996). Another is that companies are either trying to "hijack" environmentalism to serve their own commercial interests (Welford 1997) or are responding with greenwash, trying to convince consumers and shareholders that the firm is environmentally responsible while actual practices remain the same (*IDS Bulletin 1999*; Beder 1997). Another common theme is that firms are relocating in response to the raising or lowering of environmental standards, creating "pollution havens," especially in developing countries (Clapp 2002). Still another strand of the literature challenges the greenwash and pollution haven claims (Wheeler 2002), arguing that some firms appear to exceed legal environmental requirements voluntarily (Prakash 2000), sometimes even raising the overall environmental standards of developing countries—what Ronie Garcia-Johnson (2000) calls "exporting environmentalism."

This chapter demonstrates the value of embedding the analysis of business and global environmental regimes within domestic socioeconomic and political contexts. This includes both the particular features of a sector like tropical forestry as well as a developing region like the Asia-Pacific. Quite naturally, studies of global environmental governance tend to stress the norms and rules at the global level. Such studies may wrongly assume, however, that global norms and legal rules translate in similar ways in different locales. Such studies may also exaggerate the importance of formal agreements and objective interests, and underrepresent the importance of local perceptions, bargains, norms, and practices—many of which may be technically illegal. Such studies may also overgeneralize about corporate behavior and strategies in global environmental governance. The corporate strategy to support a global environmental treaty, for example, may mean something very different for a firm working in a developing country with weak enforcement than in a developed country with strong enforcement. Finally, such studies may exaggerate the effectiveness of a global regime, incorrectly assuming a new treaty or a new certification program will lead to real change to corporate conduct on the ground. That said, this chapter shows, too, that it would be a mistake to ignore international agreements or markets since corporate conduct transforms as a result of the clash of both formal rules and informal norms and perceptions, from the global to the local level. With this in mind, the chapter now turns to outline the international forests regime.

International Forests Regime

An international regime is "an institution or, more precisely, a set of norms, principles, rules and decision-making procedures that govern a particular issue area" in international relations. Regimes are "more than international organizations or even formal legal arrangements between states. They comprise the whole range of understandings, rules and procedures . . ." (Vogler 2000, 17). It is vital to be absolutely clear here that regimes are more than formal legal rules since, unlike most global environmental regimes, the international forests regime does not contain a core treaty. Defining an international regime as a "legal treaty or series of treaties that involve international commitments for policy action to address a problem," as Dimitrov (2003) does, would lead to the misleading conclusion that deforestation is a "nonregime."

The international forests regime arises from a plethora of sources, including the global discussions since the early 1990s to consider developing a global treaty for forests, the standards of organizations such as the Forest Stewardship Council, the sustainable forest principles of institutions such as the International Tropical Timber Organization, and particular provisions of international conventions directed toward other issues, such as biodiversity, desertification, climate change, and wetlands. A global treaty for forests was first seriously discussed during the 1991–1992 United Nations Conference on Environment and Development Preparatory Committees. At the 1992 UNCED Earth Summit in Rio de Janeiro, however, no binding agreement on forests was reached. Instead, the conference adopted the "Nonlegally Binding Authoritative Statement of Principles for a Global Consensus on the Management, Conservation and Sustainable Development of all Types of Forests." Rio also produced "Combating Deforestation," chapter 11 of Agenda 21. Since then, subsequent discussions over whether to proceed with developing a global convention for forests have further clarified and solidified these foundational regime principles.

The three years following UNCED saw numerous international meetings of forestry experts. The focus turned to the situation in all forest areas, rather than just in tropical zones. These meetings were partly intended to resolve differences between the North and the South, as these were seen as a key stumbling block at Rio. Following these discussions, the UN Commission on Sustainable Development created the Intergovernmental Panel on Forests (IPF) in 1995 to consider global forestry issues in greater depth. The IPF met four times over the next two years. It consistently recommended that intergovernmental policy dialogue should continue despite the failure to conclude talks on financial assistance, trade, and the feasibility and desirability of a global forests convention. The reasons the talks failed to produce consensus included continuing differences between the North and the South on market access, trade impediments, and the balance between the environmental, economic, and social goals of trade. Other difficult topics included land tenure and use, global financing for sustainable forest management, environmental technology transfers to developing countries, the role of international forest-related organizations, and the nature of decision making for forestry management (International Institute for Sustainable Development 1997, 1999). Despite the lack of full consensus, however, the IPF still managed to produce over 100 proposals for action, mainly relating to sustainable forest management issues such as the development of national forest programs, and criteria and indicators.

At the 1997 UN General Assembly Special Session (GASS)/Rio +5, discussions on forests continued to focus on whether to begin negotiations on a legally-binding international agreement on forests. Again, no consensus was reached, though importantly, Rio +5 established the Intergovernmental Forum on Forests (IFF) to continue this intergovernmental dialogue over the next few years. The Food and Agricultural Organization (FAO; 2001, 104) eventually concluded that the IFF "achieved notable progress in building consensus on international forest policy issues," narrowing the gap between the respective stances at UNCED and Rio +5. At the final meeting of the IFF in 2000, however, the intergovernmental debate was nevertheless split over whether to begin negotiations for a legally binding convention or create a United Nations Forum on Forests (UNFF) to promote existing agreements.

In the end the IFF recommended the UNFF option, although it further recommended that the UNFF should develop parameters for a legal framework covering all types of forests within five years (FAO 2001, 105–106). Later in the year the UNFF was established, with universal membership, as a subsidiary body of the Economic and Social Council of the United Nations. It is assisted by the Collaborative Partnership on Forests (CPF), established in 2001, with a mandate to "support the work of the UNFF in the promotion of the management, conservation and sustainable development of all types of forests and in the strengthening of political commitment to this end" (United Nations 2001). The ongoing work of the UNFF and CPF is based on the Rio Declaration, the Forest Principles, chapter 11 of Agenda 21 and the IPF/IFF proposals for action.

This short review of forest negotiations since Rio merely touches on some of the key trends and changes. Countless local, national, and regional discussions on forest management—connected directly and indirectly to these global negotiations—ran alongside. Numerous other international environmental negotiations and subsequent international conventions have also addressed forest management issues. The net effect of these global, regional, and national meetings and agreements was to facilitate communication across cultures and stakeholders. The process also consolidated the emerging global norms on how to manage forests sustainably. Yet there are still intense disagreements—perhaps even irreconcilable ones—among participants in global forest negotiations on the causes of and solutions to deforestation and forest degradation. Strong opposition to a global forest treaty also remains, including powerful states, international organizations, corporations, and even some NGOs.

Interstate negotiations are only one source of the rules and norms of the international forests regime. The International Tropical Timber Organization (ITTO) is one of the most important sources of management standards for tropical timber. The ITTO is the result of a legally binding commodity agreement (1983, renegotiated in 1994) for tropical timber consumers and producers, with guidelines for the sustainable management of tropical timber. Members also agreed to the year 2000 as a nonbinding target for limiting international trade to tropical timber from sustainable sources. One of the ITTO's most important tasks is to collect reliable statistics on production, consumption and trade in tropical timber. The ITTO (2003) states that it:

recognizes that a continuing supply of tropical timber on the world market depends on quality information about the trade and market place, on efficient timber production and processing methods and on sustainable forest management practices. The ITTO facilitates discussion, consultation and international cooperation on issues relating to the international trade and utilization of tropical timber and the sustainable management of its resource base.

The ITTO, however, does not have the power to enforce higher environmental standards. Gale's (1998, 3) case study of the ITTO and the tropical timber trade regime shows "that those who benefit from the current normative structure governing tropical rainforests formed a blocking coalition to prevent the regime's necessary restructuring to ensure genuinely sustainable management of tropical rainforests and the long-term, continuous supply of tropical timber."

Other international organizations, however, help to reinforce some of the stated principles of the ITTO. Lipschutz (2001, 176) argues that ". . . international forestry law does, indeed, exist. It does not, however, come in the form generally expected, as international convention and protocol. Rather, regulation is taking place through and in the market, fostered by a growing number of private and semipublic organizations and entities." The Forest Stewardship Council is one of the more important nongovernmental sources of environmental regulation of forests. It is a nonprofit organization, founded in 1993, that certifies forests as "well-managed." According to the FSC (2003), "If the forest operations are found to be in conformance with FSC standards, a certificate is issued, enabling the landowner to bring product to market as 'certified wood,' and to use FSC trademark logo." By 2002 the FSC had certified 25 million hectares of forest in fifty-four countries (Humphreys 2003).

At the heart of the international forests regime is the idea of sustainable forest management. Humphreys (1999, 251–252) writes:

The forests regime has coalesced around the core concept of sustainable forest management (SFM) and the norm that forests should be conserved and used in a sustainable manner. While an unambiguous and universally agreed definition of SFM remains elusive, the concept has been widely adopted by governmental, non-governmental, intergovernmental and business actors. SFM includes the recognition that the ecological, social, cultural and economic functions of forests should be maintained.

Other principles of the international forests regime relate to the importance of conservation, ecosystem integrity, protected areas, indigenous knowledge and values, and participation of civil society (see Humphreys 1999, 2003). Yet the international forests regime is only part—and many would argue a small part—of the explicit and implicit principles, norms, and rules that shape corporate perceptions and constrain corporate actions. National and local rules and practices reshape, dilute, and override much of the emerging international forests regime. The chapter now turns to examine the rules and norms of forest management that arise from the interaction of politics, states and firms in the Asia-Pacific.

Logging Rules and Norms in the Asia-Pacific

Politics of Timber

Politics and systemic corruption distort sustainable management plans, the collection and interpretation of data, and, most significantly, the implementation of policies in the Asia-Pacific. The informal norms that arise are in many ways far more important for shaping corporate behavior than either the international forests regime or domestic forestry laws. Senior politicians in places like the Philippines, Sarawak and Indonesia have had direct links to loggers over the last four decades. Philippine President Ferdinand Marcos (1965–1986) was one of the first national leaders to use timber concessions to appease potential challengers and reward supporters, and several of his cronies built timber empires. Senior state officials, including Defense Secretary Juan Ponce Enrile and Armed Forces Chief of Staff General Fabian Ver, were linked to illegal logging (Vitug 1993b, 16–32, 44). Bureaucrats and enforcement officers profited from logging as well, ignoring formal forest and tax rules in exchange for bribes, gifts, and personal security. Loggers in the Marcos era made windfall profits, in part by disregarding environmental and harvesting guidelines. The extent of destruction was so great that relatively little valuable commercial timber was left by the time Corazon Aquino (1986–1992) took office in 1986.

The fall of Marcos severed many of the direct ties between loggers and politicians. President Aquino did not develop timber allies during her term. Her replacement, President Fidel Ramos (1992–1998), did have some personal ties to logging, but he severed these just before taking office. Nevertheless, some loggers have managed to survive in the post-Marcos era, especially in remote areas that still contain valuable timber. Here, loggers can still wield considerable power, sometimes through a member of Congress or a provincial leader or a local mayor (Vitug 1993a, 62–68). Of 200 members of Congress elected in 1992, for example, seventeen had ties to loggers (Coronel 1996, 13). There are, as well, still members of the Philippine Department of Environment and Natural Resources with personal connections to the timber industry (Bengwayan 1999). As before, loggers work with state allies to evade forest regulations.

The politics of timber in Indonesia under President Suharto (1965/ 1967–1998) is similar to the Marcos era in the Philippines (Dauvergne 1997; Ross 2001). In the late 1960s Suharto distributed timber concessions to reward generals, appease potential opponents within the military, and supplement the military budget. The armed forces were in control of at least fourteen timber firms by 1978 (Robison 1978, 28). Through the 1980s the military's direct involvement in logging declined somewhat, although even in the 1990s there were still a few large operations. The most prominent was the International Timber Corporation of Indonesia, which was logging the biggest concession at the beginning of the 1990s (around 600,000 hectares of East Kalimantan). The armed forces owned 51 percent of this company. A conglomerate chaired by Suharto's son, Bambang Trihatmodjo, owned 34 percent. Bob Hasan, a golfing friend and major financial supporter of Suharto, owned the rest (International Timber Corporation Indonesia 1992).

In the 1980s and 1990s, five or six corporate groups controlled most of Indonesia's major timber operations (Poffenberger 1997, 456). The two most influential timber tycoons during this time were Hasan and Prajogo Pangestu. As head of the plywood cartel Apkindo from the mid-1980s until 1998, Hasan was able to dominate the timber export market by setting prices, production quotas, and export destinations for plywood manufacturers—the ban on the export of raw logs added to

Hasan's powers. Hasan's Kalimanis Group also held logging rights to 1.63 million hectares at the end of the 1990s. *Forbes* magazine estimated Hasan's assets in the mid-1990s at around U.S.$1 billion (see Barr 1998 for background). Prajogo's total forest area was even larger than Hasan's by the end of the 1990s. His firm Barito Pacific Timber controlled about 3.5 million hectares, through twenty-seven companies. Both Hasan and Prajogo relied on networks of allies within the bureaucracy, including enforcement officers and customs officials, and both maintained personal and financial links to the Suharto family.

The Indonesian Ministry of Forestry and Plantations revealed in mid-1999 that Suharto's family and allies held interests in twenty-seven timber firms and eight timber plantations, covering over 7 million hectares. Of these, members of the Suharto family personally held majority shares in twelve timber firms and four timber plantations, covering 3.8 million hectares (*Jakarta Post* 1999b). These allies and family members became vulnerable after Suharto lost power. In 1999 the government revoked eight forest concessions (covering 1.17 million hectares) granted to Suharto's children and associates, claiming these were a result of nepotism and collusion. The government further announced that it would not renew another thirteen concessions after March 2000 (covering 1.36 million hectares) (*Jakarta Post* 1999a). The government also began to investigate Hasan, Prajogo, Suharto's eldest daughter, and Suharto's half-brother for misuse of reforestation funds and corrupt forest activities. In 2001 Hasan was found guilty of corruption and sent to prison.

Today, many Suharto-era loggers still survive, including many so-called "illegal" loggers. The army, police, and allies in the bureaucratic and political realms continue to protect them. The comment of the South Kalimantan governor, responding to a question about the role of the army and police in the rise in illegal logging, is revealing: "The cooperation is clear. It's an open secret" (Greenlees 2000, 13). A large number of new loggers—both big firms and local people—have begun operations as well, some because of the opportunities opened up by more democratic structures and the decentralization of resource management to the outer islands, others because of anger over the years of repression and marginal benefits for locals of forest exploitation. The volume of illegal log production is now about 60 million cubic meters per year—more than legal production—which costs the government about U.S.$3 billion annually in tax revenue (Witular 2002).

Timber profits are even more crucial for politicians in the small and rural economies of the Asia-Pacific. In the East Malaysian state of Sarawak, for example, timber revenue is the core of the entire economy, earning roughly U.S.$1.5 billion a year in the late 1990s. Timber revenue fortifies the rule of Sarawak's chief minister, Datuk Patinggi Tan Sri Haji Abdul Taib Mahmud (1981–present). Like Marcos, Chief Minister Taib has granted forest concessions to his financial backers, friends and family (Cooke 1999; Institut Analisa Sosial 1989, 73–4; Yu Loon Ching 1987). Other senior politicians in Sarawak also rely on timber profits to win elections and retain power. Even the former minister of the environment and tourism, James Wong Kim Min, owned 55 percent of a timber firm in the mid-1990s (Limbang Trading, with a 310,000-hectare concession). The six biggest Malaysian Chinese loggers in Sarawak all maintain close ties to politicians and bureaucrats, providing financial support in exchange for licenses, political protection, and bureaucratic favors (Pura 1994). The largest logger in Sarawak is Tiong Hiew King, head of the Rimbunan Hijau Group, which controls approximately 800,000 hectares in Sarawak as well as vast areas overseas.

State Capacity

The political, military, and bureaucratic ties to loggers weaken the capacity of administrations in the Asia-Pacific to implement the principles of the international forests regime as well as national environment and forest laws. Inappropriate policies and political and social turmoil further undermine capacity. So does the lack of funds and well-trained staff. There were only about sixty employees in the Solomon Islands Forestry Division even before a coup shattered the country's administrative structure in 2000. The Philippines has only one guard for every 3,000 hectares of forest. The situation is even worse in places such as East Kalimantan, which has just one forest official for every 14,000 hectares of production forest. Only one-fifth of this staff has training in forest management. Budgets are also insufficient and staff in all of these places must rely on loggers for transportation, accommodation, and personal protection during on-site inspections.

The 1997–1999 Asian financial crisis weakened enforcement further, as firms cut even more corners to survive and rebuild, and as the collapse in the value of local currencies undercut government budgets and bureaucratic salaries. The real gross domestic product in the Solomon Islands, for example, fell by about 7 percent in 1997. Indonesia's economy contracted by almost 14 percent in 1998. In this setting there was a rise in illegal fishing, logging, mining, waste disposal, and most devastatingly, the clearing of degraded forests for plantations (especially oil palm). Fires in Indonesia, lit mostly by firms to clear land for plantations, swept out of control in the dry months of 1997 and 1998, burning as much as 10 million hectares, and covering Indonesia, Malaysia, and Singapore in a smoky haze (Barber and Schweithelm 2000). Not all of the repercussions of Asia's financial crisis, however, added to the pressures on the commercial forests. Demand for tropical timber plummeted in late 1997 and 1998 as the two main buyers, Japan and South Korea, fell into recession. As a result, Indonesia's commercial log production in 1998 was approximately 4 million cubic meters lower than two years earlier. In the Solomon Islands production dropped about 150,000 cubic meters over the same period. In both countries, however, it was still well over theoretically sustainable levels. Moreover, demand for tropical timber began to rise even before the crisis ended in 1999, partly because of China's decision to restrict domestic logging following floods. In 1999, China replaced Japan as the world's largest importer of tropical logs, and by 2001 China's total tropical log imports had reached 7.3 million cubic meters, accounting for over 40 percent of total imports of members of the International Tropical Timber Organization (ITTO 2002, 17).

Corporate Loggers

Politics and administrative capacity are not the only reasons for weak implementation of international and domestic rules of forest management. The structures and strategies of timber firms themselves pose particular problems in the Asia-Pacific. Loggers often remain highly mobile, both within and across countries, in search of cheap, high-quality stocks. Over the last four decades, Japanese buyers have traveled from the Philippines, to East Malaysia, to Indonesia, and then to Melanesia. Meanwhile, Malaysian timber firms, especially from Sarawak, have expanded to other parts of Southeast Asia, as well as Melanesia, South America, and Africa.

To help them remain mobile and competitive in the global market, many timber firms operate within large groups of affiliates and associates—including multinational groups such as Malaysia's Rimbunan Hijau, as well as domestic conglomerates like Indonesia's Kalimanis Group under Bob Hasan. (Small local sawmillers, on the other hand, generally use older equipment and often get left with the less valuable logs, leaving them unable to compete in high-end markets.) Short-term profit is the bottom line for most of these large loggers, even if this involves sacrificing the future viability of an operation in a particular location. They tend to increase production when prices and demand rise, and slow or suspend operations if prices and demand fall. In the Philippine town of San Marino, for example, during peak periods sawmills would operate twenty-four-hours-a-day. By 1990 the boom in San Marino was over. As is typical, loggers did not leave "the town even a single bridge or all-weather road across the river separating it from its vast, now denuded hinterlands" (van den Top 1998, 202).

Corporate trading groups also constrain and shape the strategies and environmental conduct of firms. Particularly significant are the chains of the Japanese general trading companies (*sogo shosha*). These firms, especially the big six—Mitsubishi Corporation, Mitsui and Company, Itochu and Company, Sumitomo Corporation, Marubeni Corporation, and Nissho Iwai Corporation—made Japan by far the world's largest consumer of tropical timber over the last thirty years. At the height of the log export booms in the Philippines, East Malaysia, Indonesia, and Melanesia, Japan imported 40–60 percent of total log production (calculated in Dauvergne 1997). Japanese general trading companies, using funds from affiliated banks in Japan, finance and facilitate links among loggers, shippers, Japanese timber processors, and even final consumers, which are often construction firms. They earn income primarily by charging commissions for services—such as for up-to-date market information, translation, and insurance against sudden fluctuations in exchange rates—as well as interest on loans. Whenever possible, these firms have tended to avoid direct investments in logging operations, preferring instead to negotiate log purchasing agreements; often, these are informal agreements, as formal ones can violate sales and export regulations. These firms, compared to Indonesian and Malaysian loggers, show less interest in trying to maximize profits at a particular point in the trade chain. Instead they focus on securing large trade volumes, partly to maintain the trade chain and partly to earn commissions on these transactions.

The funds, services, and markets of Japanese general trading companies have allowed many loggers within Southeast Asia and Melanesia, as well as tropical timber processors within Japan, to start-up and continue operations. These firms controlled almost all of the tropical timber imports into Japan in the 1960s and 1970s. The Indonesian plywood cartel, Apkindo, under the firm hand of Bob Hasan, broke ranks in the mid-1980s, deciding to do without the services of these traders, and establishing a marketing arm in Japan (Nippindo). Then, to push Japanese plywood processors out of business, Apkindo began to flood Japan with high-quality, cheap plywood in the late 1980s and 1990s. By 1993 Japan was importing over 3 million cubic meters of Apkindo's plywood. This competition drove a large number of Japanese plywood processors into bankruptcy in the 1990s. With fewer Japanese plywood processors and fewer stands of accessible commercial forests in the Asia-Pacific, Japanese trading companies began to lose interest in trading unprocessed tropical logs in the 1990s. By 1997, Japan's volume of unprocessed tropical log imports was below 6 million cubic meters, from over 13 million cubic meters a decade earlier. It had fallen even further by 2000, down to 3.1 million cubic meters (ITTO 2002).

Apkindo's tactics enabled Indonesia to pry open overseas plywood markets and expand production quickly. Plywood production was a minor industry in Indonesia in the early 1980s, but just a decade later annual production was nearly 10 million cubic meters and Apkindo was in control of three-quarters of the world trade in tropical plywood. Yet this market success came at considerable environmental cost. To open markets in countries like Japan, Apkindo lowered prices, partly relying on terrible environmental practices to keep production costs low. To sustain these market advances, Apkindo also pushed log production in Indonesia to new heights. Production was over 40 million cubic meters per year by the mid-1990s, far higher than in the 1970s when foreign firms dominated Indonesia's timber industry. Apkindo's tactics involved large economic costs as well. It charged domestic plywood processors hefty fees—for example, Apkindo collected at least U.S.$1 billion from 1983 to 1993 just from its export promotion and market development levy. Apkindo did not use these funds, however, to promote better forest management. Much of it went instead to support Hasan's business empire. The rest went to political and bureaucratic allies, including Suharto and his family.

The history of Apkindo in Indonesia shows the potential environmental drawbacks of collusive ties and financial improprieties among domestic firms. Webs of foreign and local firms in other parts of the Asia-Pacific—webs that obscure environmental transparency and responsibility—are equally problematic. Home companies, many based in Sarawak, use subsidiaries, nominees, and affiliates to conceal true ownership and control. Some of these ties revolve around ethnic or family relations, especially ethnic Chinese (including Bob Hasan, who changed his name from Kian Seng). Others arise from corporate deals, many illegal, to increase profits. The connections among timber firms are not necessarily stable, as firms make deals, withdraw, and maneuver. This can make it hard to determine the real power behind a timber firm (even the name of the owner). Intricate layers of firms enable the group to hide profits, evade taxes, and avoid responsibility for environmental damage. Moving logs through informal corporate trade chains also helps firms to conceal illegal logging and smuggling. These webs of firms can also allow foreign investors to conceal the extent of their influence from the host government. This is the case with Rimbunan Hijau's network of companies in Papua New Guinea, which Colin Filer (1997b, 215) estimates now controls half of all log exports. His method of calculation reveals the difficulty of determining the extent of a foreign logger's influence. He clustered firms, connecting them "by mutual shareholdings, overlapping directorships, or shared office facilities (as revealed by common fax and phone numbers), while their links to national companies were typically revealed by the use of sequential cheques from a single chequebook to pay their registration fees to the Forest Authority" (Filer 1997b, 213).

The nature of profits for timber firms contributes further to destructive environmental management. At times, the government fees to assess the value of a standing tree—known as the stumpage value—have totaled less than 1 percent of the eventual consumer price. The government's share of timber rents—the profits from logging after subtracting "normal" operating costs and investment returns—has been consistently low. The Indonesian government only captured 5–35 percent of timber rents from 1973–1994. Sarawak, meanwhile, captured on average less than 20 percent of rents from the mid-1960s to mid-1980s (Ross 1996, 147; Vincent 1990). There are many reasons for low revenues from forest fees: incompetence; weak state capacity to collect; policies to lure and retain overseas investors; and subsidies to prop up domestic timber processors. Deals between loggers and corrupt officials further undermine government revenues. In the Solomon Islands, for example, the Malaysian firm Integrated Forest Industries was exempted from taxes in 1995 after the firm handed out large sums to senior politicians and bureaucrats (Forests Monitor 1996, 8).

To maximize profits, timber companies tend to work within the bounds of the rules that governments and communities are able to enforce, persistently trying to bend these in their financial favor. They use, for example, an array of illegal tactics to evade fees and taxes. They transfer and hide profits and expenses through webs of affiliates and nominees to reduce corporate income taxes. In Papua New Guinea, not a single corporate logger declared a profit before 1986. Timber firms also forge export documents for volumes, species, prices and grades of log shipments to lower log export taxes (often a government bases these on the declared value of the log shipment). The Solomon Islands export records in the mid-1990s, for example, showed large differences in the average "declared" log prices, even when the species and quality appeared to be identical. Notably, firms with exemptions on log export taxes declared much higher prices (Dauvergne 1998–1999, based on data supplied to the author by government officials).

On occasion loggers simply smuggle out timber and avoid all taxes— a practice sometimes revealed by large discrepancies in import and export figures. Malaysia, for example, reported to the International Tropical Timber Organization an import figure of 700,000 cubic meters of Indonesian logs in 1996. The Indonesian government, on the other hand, reported no log exports to Malaysia in that year (ITTO 1998, 3). Governments lose considerable revenue because of tax evasion and smuggling. The Barnett Commission estimated that transfer pricing cost the government of Papua New Guinea more than U.S.$30 million during 1986 and 1987. The Solomon Islands Ministry of National Planning and Development (1999, 72) calculated that underinvoicing cost the Solomon Islands government nearly S.I.$500 million from 1990 to 1998. The World Bank estimated that the Cambodian government should have made U.S.$100 million from logging in 1996, yet it managed to collect just one-tenth of this amount.

Foreign loggers in Melanesia also exploit vague laws and divisions within landowner groups to negotiate highly favorable deals with customary landowners. To entice landowners to sign agreements, loggers tend to inflate the expectations of communities—promising new medical clinics and schools, permanent roads and bridges, and lasting and significant timber revenue. Sometimes they simply break these promises and agreements, moving on when outraged locals turn on them (see the chapters in Filer 1997a).

These environmental and financial practices have contributed to low tropical log prices and high trade volumes in the Asia-Pacific since the end of World War II. Revealingly, the prices for tropical logs have been consistently lower than the prices for temperate logs over the last fifty years (Vincent 1995, 243). Low prices and high volumes, in turn, partly explain why so much of the old-growth forests of the Asia-Pacific has ended up in Japan as cheap plywood panels to mold concrete. It has made financial sense to discard these panels after only a few uses, as it has been cheaper to buy new panels than to clean or recycle old ones (or use plastic, metal, or temperate panels).

Loggers in the Asia-Pacific, then, work within intricate corporate and sociopolitical networks. They pursue fast legal and illegal profits and leave behind environmental wastelands. As the next two sections show, so far the emerging international forests regime has contributed to significant reforms of domestic policies and institutions, yet these reforms have done little to alter the underlying sociopolitical and economic incentives and norms that actually shape corporate choices and actions.

Policy Reforms

Over the last decade or so, governments in the Asia-Pacific have passed new environmental and forestry laws. The causes and rationales for these reforms vary considerably, but many claim to address some of the failures noted previously. The Australian aid agency, AusAID, was able to push the Solomon Islands government, for example, to appoint an Australian national as commissioner of forests in the late 1990s. The Solomon Islands government also passed the *Forests Bill 1999*, replacing a myriad of confusing and contradictory laws. The Bill improved the process of determining customary land rights, created a mandatory Code of Practice for loggers, and gave the government more control over foreign timber investors. The Indonesian government also passed new forest legislation in 1999, replacing the *1967 Basic Forestry Act* with the *1999 Forestry Act*. The new Act makes loggers responsible for fires in their forest concessions unless they can provide exonerating evidence. A mix of domestic and global pressures pushed Indonesia to pass the new legislation. The World Bank, for example, temporarily suspended loans in the mid-1990s to push Indonesia to improve commercial timber management. The IMF also put in place loan conditions to require the Indonesian government to abolish most of Apkindo's monopolistic powers in 1998.

Many Asia-Pacific governments in the 1990s toughened the penalties for illegal loggers and tax evaders. Sabah, for example, amended the Forest Enactment in 1992 so illegal loggers would face jail terms of up to seven years. Sarawak followed a year later, passing new laws to stipulate a one- to five-year mandatory jail sentence for illegal loggers. In Indonesia, too, new laws mean corporate executives of firms caught clearing land illegally with fires now face long jail terms and high fines.

Governments have also set aside forests for conservation and biodiversity. Sarawak, for example, has designated over 1 million hectares of forest as "Totally Protected Area," mostly national parks and wildlife sanctuaries. This is about 12 percent of total natural forests. Indonesia now has 37 parks and 19 million hectares of conservation forest. Indonesia has also pared back and reoriented the transmigration program, which resettled people from Java to the outer islands. These reforms followed a series of environmental disasters on the outer islands, such as the peat fires that were a main source of the haze in 1997–1998. Indonesia has also revoked or refused to renew logging licenses linked to Suharto cronies. Some governments have also announced lower timber production targets, such as the Solomon Islands government's call in the late 1990s to decrease log production to 400,000 cubic meters. It is important to note, however, that in Melanesia "targets" like these are largely symbolic given that customary landowners negotiate contracts with loggers.

There is, in addition, more focus on sustainability, community forestry, and the development of nontimber products in forest management policies in the Asia-Pacific. These challenge the more conventional focus on maximizing sustainable yields and establishing commercial plantations (reforestation and afforestation). A sustainable yield is the amount loggers can harvest in a cutting cycle and still achieve equivalent future commercial volumes. The idea assumes that with "correct" selective logging techniques, tropical timber is a fully renewable resource. Selective logging requires harvesters to only remove designated species over a minimum diameter, sometimes relying on natural regeneration, and sometimes on pruning and replanting to ensure the forest returns to its original condition. Most governments in the Asia-Pacific now place "sustainable yields" and "selective logging" in the context of "sustainable forest management."

There is also some debate within the governments of the Asia-Pacific about ways to move beyond large-scale selective logging. Reduced Impact Logging (RIL) is one possibility. Under RIL, loggers harvest a small number of trees with strict controls on the minimum size of logs, the location and direction of felling, the movement of equipment, and the protection of waterways and natural ecosystems. These techniques seem to work well in at least some demonstration sites (author's field visit, Vanuatu, July 1999). Yet professional foresters tend to operate these sites and foreign aid tends to finance them. Such sites, as a result, are insulated from the pressures of the real world, such as the effects of price fluctuations, political pressures, corruption, and most significantly, timber firms pursuing fast profits. So far, there are few real-world RIL sites and most commercial logging in the tropical Asia-Pacific is still large-scale selective logging.

The most impressive policy reforms have been those to raise forest fees and taxes, giving governments a larger share of the financial rewards from logging. The Indonesian government, for example, raised the total forest fees by about 50,000 rupiah per cubic meter from 1986 to 1995 (Ross 1996, 148–150). Some governments now auction timber licenses in an effort to increase the financial value of forest stands. Some governments are also trying to inspect corporate tax claims and customs declarations more rigorously. Papua New Guinea, for example, hired the Swiss firm Société Générale de Surveillance (SGS) in the mid-1990s to monitor timber exporters. The general manager of the local SGS subsidiary claimed that it raised government revenue by over U.S.$1 million in the first six months alone (Kakas 1995).

Reforms on the Ground

These reforms have so far done little to alter the basic environmental practices of timber firms. In practice, sustainable forest management still tends to mean sustained-yield forest management (Gale 1998, 102). One exception is the Philippines, where, with little commercial timber left, alternative models have been able to gain a strong foothold. Serious technical problems also remain with sustainable forest management in the Asia-Pacific. The theory of selective logging continues to assume that forest inventories are accurate. Yet throughout the Asia-Pacific these remain speculative and highly optimistic, sometimes because of practical problems with calculations, sometimes for political reasons. Governments in the region will, on occasion, abruptly change the estimate of sustainable yields for political rather than scientific reasons. The concept of a sustainable yield becomes particularly problematic when governments use it to measure and manage forests on a provincial or national scale. Indonesia's national sustainable yield, for example, is still commonly put at around 20 million cubic meters per year. Such a measure in effect justifies heavy logging in one area (e.g., Kalimantan) based on the logic that another area (e.g., Irian Jaya) is left untouched, even though the areas may not have equivalent social, economic, or biological values.

Calculations of sustainable yields continue to assume, too, that loggers leave forests in reasonable environmental condition. Study after study shows, however, that loggers rarely do—in Papua New Guinea, for example, in some cases 70 percent of the trees are "mortally wounded" after selective logging (Millett 1997, 325). Despite such studies, it is still common for governments, even NGOs, to call for firms to lower harvests to some hypothetically "sustainable" level in the pursuit of "sustainable forest management." The model of sustainable forest management in the Asia-Pacific, then, continues to do more to legitimize corporate activities and reinforce corporate values than protect the environment. Furthermore, it continues to provide a potent linguistic tool to dismiss alternative views. Indigenous peoples like the Penan of Sarawak are still treated as backward and primitive. It also gives loggers a "scientific" language to deflect critics and delay reforms. Loggers and state allies, in response to charges that selective logging is irreparably damaging the forests, are also able to divert the broader debate at global and regional meetings toward ways of honing harvesting techniques and management models.

Particular flaws with the concepts of "sustainable forest management" are only a small part of the problem, however. Most loggers do not in fact work within the parameters of these formal management models, but rather within a set of informal rules and norms. Loggers continue to ignore preharvest surveys and preparation as well as postharvest silvicultural treatments. They continue to harvest trees outside of legal concessions, damage or cut undersized and protected trees within the legal concessions, and construct roads and bridges with little regard for the environmental or social implications. They continue to pollute rivers and streams, sometimes creating pools of stagnant water that spread malaria. They continue to log sacred sites and wildlife habitats. And most continue to ignore reforestation or silvicultural responsibilities. In the Melanesian countries of Papua New Guinea and the Solomon Islands, once loggers gain access, they also continue to break agreements with landowners, especially verbal promises (Dauvergne 2001).

An environmental sketch of the region paints a bleak picture. The Solomon Islands *1995 Forestry Review* (1995) claimed logging practices in the Solomon Islands "are amongst the worst in the world." The practices of loggers in Indonesia, Sarawak and Papua New Guinea are not much better. A review of loggers during the Suharto era found that just 4 percent were responsible managers (Potter 1993, 113)—if anything, the situation has deteriorated since then. A study in Sarawak found that bulldozers typically traverse 30–40 percent of a concession. Loggers damage 40–70 percent of the trees by the time they are finished, leaving no real chance that natural regeneration will restore the ecosystem. Some loggers in Sarawak also reenter their concession before the end of the official cutting cycle, already a relatively short twenty-five to thirty years—many foresters estimate that it takes at least sixty years before stands can produce similar commercial harvests (Arentz 1996).

Such logging practices are rapidly destroying the tropical forests of the Asia-Pacific. Today forests only cover three-quarters of Melanesia, about half of Indonesia, Laos and Cambodia, less than half of Malaysia and Burma, less than one-quarter of Thailand, and less than one-fifth of the Philippines. Almost 95 percent of Asia's frontier forests, defined by the World Resources Institute as large and pristine enough to still retain full biodiversity, are now gone. The Philippines no longer has frontier forests, and Vietnam, Laos, Thailand, and Burma will soon follow. Cambodia has just 10 percent, Malaysia just 15 percent, Indonesia 25 percent, and Papua New Guinea 40 percent (World Resource Institute 1997).

Deforestation continues at an alarming rate. In Indonesia, for example, despite the reforms to forest management policies in the 1990s, annual deforestation has jumped to between 1.7 and 2.4 million hectares, far higher than in the 1980s and first half of the 1990s. Selective logging continues to be a key reason for these high rates of deforestation. Other forces are farmers, developers, plantation owners, and forest fires. Selective loggers are particularly significant, however, as they begin the process of deforestation by opening up the forests to these other forces. These loggers have already harvested much of the Asia-Pacific. The Philippines, as a net importer of timber, is in the worst condition, but the rest of the Asia-Pacific is not far behind. At recent rates and under current practices, loggers will largely deplete the old-growth forests of commercial timber in East Malaysia (Sabah and Sarawak) within ten years, the Solomon Islands and Papua New Guinea within fifteen years, and Indonesia within fifteen–twenty years (Dauvergne 2001).

Conclusions

The informal and political nature of state-business relations in the Asia-Pacific shapes loggers' reactions to both international and national pressures for environmental reforms. Political patrons continue to help loggers gain and maintain access to valuable commercial forests. In pursuit of high profits, both domestic and foreign firms continue to rely on networks of state and local allies to skirt regulations—such as forging export documents or smuggling logs overseas. The acceptable rules here are not always easy to discern, and can change as local leaders change. While a firm occasionally blunders and is expelled from a region, many firms appear willing to take this risk, if necessary packing up and moving to another location, often within the same country (Filer 1997a).

On a rhetorical level, firms, and even more commonly their state allies, are challenging the critics of logging practices, claiming practices are now "sustainable" or "green." Sarawak's Chief Minister Taib, for example, claims that his is now "a model state for other countries to emulate" (quoted in World Rainforest Movement 2000). At the same time, timber firms are increasingly using the language of sustainability and corporate responsibility. If pushed, corporate executives appear ready to concede to reforms to national forest policies, knowing most states are unable in the near-term to enforce them. Of course in some cases, like higher forest fees and export taxes, firms are simply unable to prevent reforms. Generally, however, firms have been able to resist and deflect efforts to impose far-reaching environmental reforms to logging practices.

This process of reform is widening the disjuncture between the rules and language from above—the international forests regime and national policies—and norms and enforceable rules below. Loggers gain some legitimacy from participation and compromise in high-level meetings, while continuing to log frantically on the ground, paying off powerful state and local allies to maintain access for as long as possible, and distributing largesse along the chain of timber production. In some cases the links are through formal industry associations (e.g., Apkindo in Indonesia), although often these are informal (such as between Japanese financiers, log buyers, processors and construction companies). Firms pay far less attention to NGOs and local organizations, as these are weak in much of the tropical Asia-Pacific and tend to exert little real pressure on loggers.

There are a few exceptions. The NGO Global Witness has exposed corruption and illegal loggers in Cambodia. Backed by donors from Britain and Australia, Global Witness was appointed as an independent forest monitor in Cambodia in 1999, reporting directly to the Hun Sen cabinet as well as to donors (Global Witness 1999). Yet most NGOs in the region have small budgets, few staff, and limited influence. Consumers in Northeast Asia also tend to be less interested than Europeans or North Americans in overseas environmental issues and in supporting environmental boycotts. Censorship, weak political oppositions, and authoritarian governments have further limited the reach of activists in the Asia-Pacific. Some timber investors, such as Malaysian firms in Papua New Guinea, however, do work hard to accommodate locals with promises of "development," in the form of medical clinics, roads, and schools, and with gifts and money.

The World Commission on Environment and Development (1987, 43) defined sustainable development as "development that meets the needs of the present without compromising the ability of future generations to meet their own needs." This definition, now standard in much of the world, including in the Asia-Pacific, requires governments and firms to plan far into the future. To promote sustainable management, logging contracts in the Asia-Pacific may guarantee a firm access to a site for twenty, fifty, even eighty years. On the surface this is logical and sensible, as a firm needs such guarantees to create the incentives to pursue sustainable yields—that is, to manage the forest so the next harvest in fifty years or so yields an equal volume of commercial logs. Yet, as this chapter shows, loggers know that access to a site often ends without notice, on the whim of a politician, bureaucrat, or military officer. Loggers, furthermore, have little confidence in the future of markets for tropical timber. Prices historically have been unstable and unpredictable. With such high uncertainty, what rational corporate executive would plan a generation ahead? A rational executive would do exactly what so many now do: log as fast as possible, placating and accommodating for as long as possible their networks of sociopolitical allies to evade costly regulations and taxes in an effort to maximize immediate profits.

The international forests regime and new environmental and forestry policies in the Asia-Pacific are no doubt gradually increasing the pressure on loggers to adapt and alter actual practices. Yet the nature of corporate resistance and accommodation means the process of change is currently far too slow to save the remaining old-growth commercial forests.

References

Arentz, F. (1996). Forestry and politics in Sarawak: The experience of the Penan. In R. Howitt (Ed. with John Connell & Philip Hirsch), *Resources, nations and indigenous peoples: Case studies from Australasia, Melanesia and Southeast Asia* (pp. 202–211). Melbourne: Oxford University Press.

Barber, C. V., & Schweithelm, J. (2000). *Trial by fire: Forest fires and forestry policy in Indonesia's era of crisis and reform*. Washington, D.C.: World Resources Institute.

Barr, C. M. (1998, April). Bob Hasan, the rise of Apkindo, and the shifting dynamics of control in Indonesia's timber sector. *Indonesia*, 65, 1–36.

Beder, S. (1997). *Global spin: The corporate assault on environmentalism*. Melbourne: Scribe Publications.

Bengwayan, M. (1999, October 11). Philippines drops total logging ban. *Environment News Service*.

Clapp, J. (2002). What the pollution havens debate overlooks. *Global Environmental Politics*, 2(2), 11–19.

Cooke, F. M. (1999). *The challenge of sustainable forests: Forest resource policy in Malaysia, 1970–1995*. Sydney: Allen and Unwin.

Coronel, S. (1996). Unnatural disasters. In S. Coronel (Ed.), *Patrimony: Six case studies on local politics and environment in the Philippines* (pp. 7–19). Pasig City: Philippine Center for Investigative Journalism.

Dauvergne, P. (1997). *Shadows in the forest: Japan and the politics of timber in Southeast Asia*. Cambridge, MA: MIT Press.

Dauvergne, P. (1998–1999). Corporate power in the forests of the Solomon Islands. *Pacific Affairs*, 71(4), 524–546.

Dauvergne, P. (2001). *Loggers and degradation in the Asia-Pacific: Corporations and environmental management*. Cambridge, UK: Cambridge University Press.

Dimitrov, R. S. (2003). Knowledge, power, and interests in environmental regime formation. *International Studies Quarterly, 47*(1), 123–150.

FAO (Food and Agriculture Organization) (2001). *State of the world's forests 2001.* Rome: FAO Forestry Publications and Information Unit, available at <http://www.fao.org/forestry/FO/SOFO/SOFO2001/publ-e.stm>.

Filer, C. (Ed.) (1997a). *The political economy of forest management in Papua New Guinea.* National Research Institute (Monograph 32). London: International Institute for Environment and Development.

Filer, C. (1997b). A statistical profile of Papua New Guinea's log export industry. In C. Filer (Ed.), *The political economy of forest management in Papua New Guinea* (pp. 207–248).

Forests Monitor (1996). Kumpulan Emas Berhad and its involvement in the Solomon Islands. Draft briefing document. Forests Monitor. April.

FSC (Forest Stewardship Council) (2003). Available at <http://www.fscoax.org/>.

Gale, F. (1998). *The tropical timber trade regime.* London and New York: MacMillan Press and St. Martin's Press.

Garcia-Johnson, R. (2000). *Exporting environmentalism: U.S. multinational chemical corporations in Brazil and Mexico.* Cambridge, MA: MIT Press.

Global Witness (1999). Global Witness appointed independent monitor of forestry. Press Release. 2 December. Available at <http://www.globalwitness.org/press_releases/display2.php?id=57>.

Greenlees, D. (2000, July). $8.6 bn graft cuts Indonesian forests. *The Australian,* 22–23, 13.

Humphreys, D. (1999). The evolving forests regime. *Global Environmental Change, 9*(3), 251–254.

Humphreys, D. (2003). Life protective or carcinogenic challenge? Global forests governance under advanced capitalism. *Global Environmental Politics, 3*(2), 40–55.

IDS Bulletin (1999). Globalisation and the governance of the environment. *IDS Bulletin, 30*(3), July.

IISD (International Institute for Sustainable Development) (1997, August 8). *Countdown Forests '97.* Winnipeg, Canada: IISD. Available at <http://iisd1.iisd.ca/forests/countdown/cf-8.pdf>.

IISD (1999). Trend watch: International forest dialogue. *Global Vision for Forests 17* March/April. Available at <http://iisd.ca/didigest/mar99/mar99.1.htm>.

Institut Analisa Sosial (1989). *Logging against the natives of Sarawak.* Selangor, Malaysia: INSAN [Institut Analisa Sosial].

International Timber Corporation Indonesia (ITCI). (1992). *PT. International Timber Corporation Indonesia.* Jakarta: ITCI.

ITTO (International Tropical Timber Organization) (1998). *Annual review and assessment of the world tropical timber situation: 1997.* Yokohama: ITTO.

ITTO (2002). *Annual review and assessment of the world timber situation: 2001.* Yokohama: ITTO.

ITTO (2003). Available at <http://www.itto.or.jp>.

Jakarta Post (1999a, July 9). Government revokes vast forest concessions.

Jakarta Post (1999b, June 23). Soeharto's control 7.15 mha of forests in the country.

Kakas, D. (1995, August 19). Export checks increase forest revenue, SGS predicts further revenue rise. *The Saturday Independent.*

Lipschutz, R. D. (2001). Why is there no international forestry law? An examination of international forestry regulation, both public and private. *UCLA Journal of Environmental Law & Policy, 19*(1), 155–182.

Millett, J. (1997). The economics of sustainable development in Papua New Guinea. In C. Filer (Ed.), *The political economy of forest management in Papua New Guinea* (pp. 311–332).

Paterson, M. (2001). Risky business: Insurance companies in global warming politics. *Global Environmental Politics, 1*(4), 18–42.

Poffenberger, M. (1997). Rethinking Indonesian forest policy: Beyond the timber barons. *Asian Survey, XXXVII*(5), 453–469.

Potter, L. (1993). The onslaught on the forests in South-East Asia. In H. Brookfield & Y. Byron (Eds.), *South-East Asia's environmental future: The search for sustainability* (pp. 103–123). Tokyo and Melbourne: UN University Press and Oxford University Press.

Prakash, A. (2000). *Greening the firm: The politics of corporate environmentalism.* Cambridge, UK: Cambridge University Press.

Pura, R. (1994, February 15). Timber baron emerges from the woods. *Asian Wall Street Journal, 1,* 4.

Robison, R. (1978). Toward a class analysis of the Indonesian military bureaucratic state. *Indonesia, 25,* 17–39.

Ross, M. L. (1996). *The political economy of boom-and-bust logging in Indonesia, the Philippines, and East Malaysia: 1950–1994.* Unpublished doctoral dissertation, Princeton University.

Ross, M. L. (2001). *Timber booms and institutional breakdown in Southeast Asia.* Cambridge, UK: Cambridge University Press.

Rowell, A. (1996). *Backlash: Global subversion of the environmental movement.* London: Routledge.

Rowlands, I. (2000). Beauty and the beast? BP's and Exxon's positions on global climate change. *Environment and Planning C: Government and Policy, 18*(3), 339–354.

Solomon Islands 1995 Forestry review (1995). An unofficial internal government document. Unpublished manuscript. Supplied by a government official during the author's field research in the Solomon Islands.

Solomon Islands Ministry of National Planning and Development (1999). *Medium term development strategy 1999–2001*. Draft. Honiara: Solomon Islands Government.

UN (United Nations) (2001, May). *Collaborative partnership on forests (CPF) policy document*. New York: United Nations. Available at <http://www.un.org/esa/sustdev/cpfpolicy.pdf>.

van den Top, G. (1998). Deforestation of the Northern Sierra Madre. In V. T. King (Ed.), *Environmental challenges in South-East Asia* (pp. 193–230). Richmond: Curzon Press.

Vincent, J. R. (1990). Rent capture and the feasibility of tropical forest management. *Land Economics*, 66(2), 212–223.

Vincent, J. R. (1995). Timber trade, economics, and tropical forest management. In R. B. Primack & T. E. Lovejoy (Eds.), *Ecology, conservation, and management of Southeast Asian rainforests* (pp. 241–261). New Haven, CT: Yale University Press.

Vitug, M. D. (1993a). Is there a logger in the house? In E. Gamalinda & S. Coronel (Eds.), *Saving the earth: The Philippine experience* (3rd ed., pp. 59–68). Pasig City: Philippine Center for Investigative Journalism.

Vitug, M. D. (1993b). *Power from the forest: The politics of logging*. Pasig City: Philippine Center for Investigative Journalism.

Vogler, J. (2000). *The global commons: Environmental and technological governance* (2nd ed.). New York: John Wiley & Sons.

Welford, R. (1997). *Hijacking environmentalism: Corporate responses to sustainable development*. London: Earthscan.

Wheeler, D. (2002). Beyond pollution havens. *Global Environmental Politics*, 2(2), 1–10.

Witular, R. A. (2002, July 4). RI, China to team up to curb illegal logging. *The Jakarta Post*.

World Commission on Environment and Development (WCED) (1987). *Our common future*. Oxford: Oxford University Press.

World Rainforest Movement (2000, March). Malaysia: How to destroy forests in Sarawak by planting trees. *Bulletin, 32*. Available at <http://www.wrm.org.uy/bulletin/32/Malaysia.html>.

WRI (World Resource Institute) (1997). *The last frontier forests: Ecosystems and economies on the edge*. Washington, DC: WRI.

Yu Loon Ching (1987). *Sarawak: The plot that failed 10 March 87–7 April 87* [A collection of newspaper articles]. Singapore: Summer Times.

8

Environmental and Business Lobbying Alliances in Europe: Learning from Washington?

David Coen

In studying environmental politics, great emphasis is placed by international relations theory on the intergovernmental negotiations at global summits and international organisations, but the impact of societal actors, such as business and consumers, are often overlooked in the day-to-day environmental governance process. This chapter explores the political reality of environmental public policymaking at the European Union institutional level and the mechanisms by which large firms influence the agenda-setting and policy formulation process in Brussels and member states.

The gradual transfer of regulatory functions from member states to the EU institutions in areas such as product quality, health and safety, employment and competition law, and environmental standards have all contributed to the Europeanization of environmental interest groups (Grant, Mathews, and Newell 2000; Sbragia 2000). However, while environmentalists are mobilised at the European level, we recognize that business has a strategic resource advantage in lobbying and takes a prominent role in both formulation and implementation of EU environmental directives (Grant 2000; Jordon 2002). Environmental groups recognize these potential structural problems of influence, but are also aware of their potential agenda-setting functions (Mazey and Richardson 2001). As a result, many are involved in complex multilevel advocacy coalitions with business and other public policy interests (Coen 1998; Sabatier 1998; Young and Wallace 2000).

This chapter contributes to the environmental governance debate by exploring the development of a complex "elite pluralism" that favors business interests in EU policy formulation, and compares this new business-government relationship with the recent U.S. experience. Mini-case studies will be used in order to explore how the EU environmental interests have adapted to the multilevel lobbying opportunity structure. In so doing, the chapter will illustrate how international business and societal interests have recognized the importance of contestation and compromises in environmental policymaking in the European Union and United States. The result helps explain the harmonization of international public policy styles around policy forums on global issues, such as climate change and the ozone layer (Levy and Newell 2000), but conversely, the chapter also illustrates how cultural and political variance allows for differences in implementation of policy in member states, and market creation and market access issues within Europe (Jordan 2002).

Multilevel European Policy Process: Venue Shopping and Feedback Lobbying

Traditionally, business interest representation in the EU has been analyzed in vertical sector terms, with national trade associations' positions feeding into European federations and European institutions (Greenwood 1997; Greenwood and Aspinwall 1998; Streeck and Schmitter 1991). However, in recent years, complex issue networks have evolved and a desire for more horizontal European interest communities has been expressed by the European Institutions (Coen and Dannreuther 2003; Richardson 2000). This desire was formalized in the recent European Commission Green Paper on Governance, which explicitly called for horizontal alliances between consumers, business and European societal interests (European Commission 2001). The result has been the birth of complex multilevel and institutional advocacy coalitions, ad hoc interest groupings, and EU institutionally led forums. In such a complex environment, no industrial or societal group can lobby in a political vacuum, and firms wishing to lobby directly on environmental issues have had to incorporate consumer demands and Green lobby positions into their political strategies (Grant et al. 2000; Young and Wallace 2000).

Why the change in public policy approaches? In policymaking terms, EU institutions, faced with a boom in public interest lobbying in the

1990s, recognized a need for some form of regulated representation, if information flows were to be managed. The Commission's informal solution has been to create policymaking forums and Select Committees (Coen 1997; Richardson 2001) based around policy insiders and policy outsiders (Broscheid and Coen 2003). Business, in turn, has recognized that to access these restricted entry policy forums, it must broaden its political and information legitimacy by being more representative of economic and societal interests. The result has been the explosion in the late 1990s of short-lived and issue-specific political alliances (Coen 1998; Richardson 2001; Webster 2002).

While EU institutions have become significant policy actors, the degree of activity in Brussels is still a function of the policy cycle, with interests focusing on agenda-setting and formulation of EU directives at the European institutions, and the implementation of directives and "day-to-day" regulatory monitoring in the member states. What is more, as we move along the policy cycle and assess different policy areas, we can expect feedback loops between the national and European institutions. For example, in the post-Amsterdam Treaty the European Parliament (EP) has increased its role in revisions of Commission policy proposals and has codecision powers with the Council of Ministers. As a result, civic and social interests have increased their voice in the policy process, and business has found a secondary channel to influence formulation of EU directives. Likewise, agenda-setting and policymaking oscillates between national and European channels, depending on whether the issue is a regulatory, redistributive, or distributive question, and thus how far it impinges on the central questions of sovereignty and subsidiarity.

As a result, we can no longer see Europeanization of business lobbying in terms of "bottom-up" management or "top-down" coordination but as a managed multilevel process with numerous feedback loops and entry points constrained by the size of the firm, lobbying budgets and the nature of the policy area (Coen and Dannreuther 2003). With regard to nature of policy, emphasis on the intergovernmental or multilevel approach is still a function of the type of policy under discussion, that is to say, the degree to which a policy is regulatory, distributive and redistributive (Coen 1998, Richardson 2001, Wallace and Wallace 2000). Hence, on regulatory issues such as environmental policy, the

European Commission can be seen to be taking a policy lead, but member states show a great reluctance to hand over redistributive powers such as taxation to the supranational level.

Recognizing the difficulty in generalizing business lobbying characteristics, some general trends can be observed since the Single European Act. First, we have observed increases in the direct business lobbying of the European institutions, and particularly at the European Commission as it has expanded its regulatory competencies (Pollack 1997). In fact, the Department of Industry (DTI) estimated that 70 percent of legislation affecting British business now emanates from Brussels (Grant 2000). However, this direct lobbying of the Commission must be seen in the context of a multichannel, multilevel lobbying strategy, as it has become accepted practice by European affairs directors that collective and direct strategies, national and European mobilization are all simultaneously required, if influence is to be maximized in the Brussels arena (Coen 1997, 1998). Moreover the strong showing of national authorities and, to some extent, national trade associations in recent years (Greenwood 2003) can be attributed to the technical standard setting (Egan 2001) and variance in national regulatory monitoring and control, in line with subsidiarity (Coen and Heritier 2000).

Recognizing policy and national variance, this chapter predominantly concentrates on the distributive and regulatory aspects of EU environmental policy. Here the EU has been seen to be proactive in agenda-setting and formulation in the 1990s and has enacted over 700 environmental laws (McCormick 2001). More significantly for the mobilization of interest representation and governance is the fact that all current policymaking must take into account environmental dimensions via the concept of "environmental mainstreaming." The growing importance of the EU level can be attributed to a number of factors, such as the rise of the Green movement in Western Europe, acceptance of the need for cross border collaboration on environmental problems, and the removal of nontariff barriers in the Single Market (Jones 2001).

However, environmental policymaking at the EU level is not a level playing field, with northern European countries such as Germany and Sweden setting higher environmental standards than Southern countries such as Greece, Portugal, and Spain. Hence we observe a tendency towards flexible and poor implementation, and recognition of minimum standards (Jordan 2002). This has changed slightly in recent years, as the Commission has attempted to benchmark member states on style and level of implementation and has become more willing to refer member states to the European Court of Justice (ECJ). Moreover, the Commission's 1999 report showed that in 1997, it referred thirty-seven cases to the ECJ and sent sixty-nine reasoned opinions to member states. Nevertheless, by 2001 almost one-third of all infringement proceedings were linked to environmental policy (Jones 2001). Accordingly it has become important that all interested parties attempt to influence the EU institutions in their interpretations and formulation of directives.

As the preceding illustrates, producer lobbies are active and exercise a stronger influence over policy formulation and agendas than traditional consumer and environmental groups (Grant 2000; McCormick 2001). Nevertheless, as the cases in this chapter demonstrate, we see different groups exercising greater influence over different institutions. The key institutions in policy formulation are the Commission, member states and, increasingly, the European Parliament. In most cases the Commission has become the focal point on the environmental directives for both environmental and industrial lobbies. However, we would expect closer ties for the environmental groups to the Environmental directorate and closer industrial ties with Enterprise, Single Market, and Competition directorates (Richardson 2001). Likewise, Environmental and consumer groups have developed stronger links, than business, to the EP—an institution that has often taken a greener standpoint (Earnshaw and Wood 1999; Sbragia 2000).

The various political lobbies have also experienced different potential alliances with member states; for example, in the emissions case presented next, the Portuguese and Spanish governments aligned with the petroleum and automobile lobbies while the British, Danes, and Germans, in line with the EP and European Commission, aligned with consumer and environmental groups. However, while alliances are often shifting between levels, countries, institutions, and even within institutions, some hard truths about access to the policy process for the key interests are evident within this policy regime.

Changing EU Architecture: The Case of the End of Life Automobile Directive

The European Commission, faced with a European Council resolution on waste management in 1990 and the EP's 1992 recycling proposals, initiated the policy on the End of Life vehicles debate in 1994. Nonetheless, while the Commission is responsible for proposing responses to problems, it has never had a monopoly on agenda-setting, influenced as it is by the Council of Ministers, EP suggestions, the Court of Justice rulings, and ultimately pressure from organized policy communities (McCormick 2001). Thus, the negotiation of the final 2000 directive on liability for the ecological disposal of cars represented an example of how business adapted to the policy cycle through consultation and formulation, and successfully managed changing EU institutional arrangements.

The Commission, acting as a political opportunist, brokered its initial proposals in 1997 and placed a great emphasis on the liability of manufacturers for the recycling and disposal of vehicles (Tenbucken 2002). While the European Automobile Trade Association (ACEA) appeared on side during early directive drafts from 1994 to 1997, the industry awoke and started aggressively lobbying against the directive, as the full cost to the industry became clear in 1998. In this period, the firms mobilized both the ACEA and directly lobbied the Commission, arguing that the directive should only apply to new vehicles. However, the Commission was unreceptive and sought to push the directive through. Recognizing that the EP first reading appeared to favor the directive, Volkswagen and other German manufactures altered their lobbying focus and mobilized national support at the regional level and placed pressure on Schroeder's government at the federal level by highlighting the cost to the German car industry. The result of such domestic pressure was that the German Environmental Minister cancelled the proposed Council of Ministers Environmental meeting in early 1999 and slowed the policy-formulation down. This strategy visibly illustrated how the national route was still a credible policy option in setting and reformulating EU directives (Tenbucken 2000).

With the successful blocking of the original proposals, radical changes were proposed by the German car manufacturers with regard to liability and recycling of heavy metals. Moreover, the extra time won at the Council of Ministers allowed firms to focus their attention on the newly empowered post-Amsterdam European Parliament. Drawing on the codecision procedures, business was able to convince the EP to call for some forty-three amendments to the 1997 Green Paper. Significantly, while focusing on the EU institutions throughout much of 1999, the German manufacturers also lobbied national governments via their subsidiaries in Spain (Seat) and the UK (Rover) to support the German government's revisions at the Council of Ministers. At the same time the German government negotiated issues linkages on fishing policy with the Spanish government and harmonization on art dealing with the British. Finally, the EP and European Commission working with the ACEA agreed on some thirty-three amendments, and the Council of Ministers accepted a Joint Text in 2000.

As the preceding discussion briefly illustrates, a successful lobby requires a number of vertical and horizontal strategies. The players must be aware of where a policy is initiated, what the alternative pressure points are, and who has the potential veto points in the process. However, having recognized that venue shopping is a coherent and viable strategy, it is important to recognize how to access the various institutions and political channels along the policy cycle. Furthermore, the preceding case shows the duality of the EU policy process, in so far as firms must develop an EU business-government identity while maintaining a strong national voice.

European Level Forums and Coordinated Environmental Issues Networks

Contemporary policy forums include many of the largest firms in Europe, suggesting the development of an inner core of policy makers and the institutionalization of big business in the EU policy process. Also, many of the European policy forums, such as the Environmental forum of DGXI, have been reinforced and guided by the success of groups including the European Round Table (ERT) and American Chamber of Commerce (AmCham). Firms and the Commission recognized early that firm-based groups such as AmCham, which drew on its U.S. lobbying experience, could provide early and detailed information. Thus, while new industry forums continued to pursue collective EU agendas, they benefited from a smaller membership of like-minded policy actors, with significant payoffs (discussed later). Issues that could no longer be resolved at a collective association level were effectively tendered out into new ad hoc political alliances that grew around single issues. In light of these developments, the voices of small and medium enterprises (SMEs) and NGOs were potentially marginalized, hence the European Governance Paper that actively called for suggestions to facilitate the inclusion of these interests (European Commission 2001).

The benefits for big business of forum politics were more than simple access. It also raised the influence of business in the power politics of inter-Director General rivalry, and has given them quasi-policymaking and agenda-setting status in certain strategic areas (Richardson 2001). While pressure for new alliances and political groupings came from the increasingly disaggregated political goals of large firms, it was also significant that competition between Director Generals (DGs) encouraged the creation of forums and networks. The advantage of specialist forums, in addition to the focused policymaking and ability of the Commission to demand specific access criteria, was that it provided the individual Commissioners with their own political and economic constituencies within Brussels and vis-à-vis member states (Broscheid and Coen 2003). The most visible recent forums have been Liikanen's Enterprise and Innovation groups on EU enterprise, competitiveness, growth, and employment and Prodi's e-Europe initiative.

New Informal Business Coalitions and Alliances

While the greatest lobbying benefits to business come via the formal Commission-led forums, it is possible to observe a secondary trend towards formalized business groups taking the policy lead in a form of public-private policymaking. A visible example of such a grouping is the Trans-Atlantic Business Dialogue (TABD), set up in 1995 as a joint initiative of the Commission and U.S. State Department to circumvent trade issue problems at the WTO. Gradually it has evolved into a quasi-policymaking organization that fast-tracks business-led trade and product standards to the EU and U.S. regulatory bodies (Coen and Grant

2001; Cowles 2001). The European Round Table has also reasserted itself in the integration process, in light of the Lisbon 2000 summit Declaration. Here the European Council leaders committed themselves to the ambitious goal of making the EU the most competitive and dynamic knowledge-based economy in the world by 2010. This business-friendly agenda explicitly recognized the importance of entrepreneurs as a means of growth and job creation, and attempted to create an innovative environment through reduction of compliance costs and coordination of regulation within the internal market. Significantly, in line with the preceding, the European Round Table (ERT) and the Union of Industrial and Employers' Confederation of Europe (UNICE) agreed to work with the Commission to develop relevant benchmarks applicable across the EU.

Such institutionalized big business representation has advantages in terms of policy delivery and the credibility of actors involved in the policy process. Notably, the Commission enjoys credibility while the large firms gain access—both sides gain in a relatively equal partnership. Central to this arrangement has been the Commission's ability to dictate terms for access and the ability of big business and their representatives to make the required changes in their behavior to win the prize of privilege in access, influence and agenda-setting powers. Clearly, business had to change its behavior in response to the rules of the multilevel game described in the previous section. But to enjoy benefits of insider status, big business had to win the trust of the Commission by becoming European in its business-government identity as well.

Faced with Commission forums, the advantages of big business, and the increased costs of lobbying, pressures on societal interests to merge and coordinate have been irresistible. Reflecting Olsonian problems of coordinating larger and less focused formalized groups, the public interest solution has been the creation of short-life issue networks that form and disband around a single focused directive—as illustrated by the case that follows. Paradoxically big business has also facilitated the creation of many of these new ad hoc alliances as they too, seek alliances with civic groups and critical mass to facilitate credibility and access to the EU policy forums.

Green Coalition Building: The Case of the Automobile Emissions Standards

In defining the directive on future emissions, the Commission initially brought together technical experts from European federations representing the car manufacturers (ACEA) and the European Petroleum Industry Association (EUROPIA) to form an insider group of experts. The aim of this initial dialogue (1993–1994) was to assess the most effective package of measures, including vehicle, technology, fuel quality, and nontechnical measures that would reduce emissions. However, the subsequent development of the European Campaign for Clear Air (ECCA) and the Auto I Program were notable in that they demonstrated how public interests can mobilize and collaborate on a single issue (Webster 2002; Young and Wallace 2002).

In response to the previously described technical meetings, new public interest coalitions formed as countervailing groups to create a green voice in the debate, the most notable being ECCA. This group had at its core six lobbyists: the European Bureau of Consumer Unions (BEUC), the Confederation of Family Organizations in European Community (COFACE), The Euro Citizen Action Service (ECAS), the European Environmental Bureau (EEB), the European Public Health Alliance (EPHA) and the European Federation for Transport and Environment (T&E). Significantly, the group saw itself as a short-life and issue-driven coalition for clear air and emissions (Grant et al. 2000). While there were differences in the aims of consumer groups, citizens, and environmental groups, they found common ground on environmental questions. Thus the ECCA gave them an official name and critical mass vis-à-vis the entrenched producer lobbies in the industrial directorate.

While highly visible, the ECCA had no formal secretariat or fixed financial contributions, unlike NGOs or conventional industry associations; rather, responsibilities were allocated according to the expertise of the member associations. For example, the ECB focused on the Environmental Directorate while the consumer groups attempted to lobby the Industrial and Single Market Directorates (Webster 2002). The result was "institutional shopping" by a flexible, focused, and fast-moving specialist political coalition. However, in addition to lobbying the EU institutions directly on technical questions, the ECCA also attempted to broaden the public policy debate on emissions via a series of conferences that brought together a wider epistemic policy community of academics, national government officials and environmental consultants. The preceding activity illustrated that complex multilevel advocacy coalitions could be fostered as environmental policy norms were established in the 1990s. However, while actively lobbying as a coalition and broadening the scope of the environmental debate, all the groups maintained individual identities and positions relating to potential countervailing issues.

By the time of the 1995 draft proposal to EFEG, two distinct camps could be identified. The first, taking a strict approach, included the Commission, Austrian, German, British, and Dutch governments, the European Parliament, environmental, and consumer groups. The second group was more permissive and included the automobile and petroleum industries, and the Spanish and Portuguese governments. The public-interest group attempted to influence the policy process with an appeal to the accountability and legitimacy of EU policy, by bringing a range of public interests together at the European, national, and local levels. Significantly, the Commission seemed increasingly receptive to the environmental voice and green taxes after the strong showing of environmental parties in national and European elections. The producer groups relied on technocratic policymaking and placed the greatest emphasis on modelling air quality and cost-benefits of changes to engine specification. Therefore, each lobbying coalition utilised its comparative advantage in "EU resources dependency" terms.

While the Commission clearly altered its general position towards the environmental lobby at the expense of the automobile industry, its proposals were not as far reaching as the 1995 proposals (Young and Wallace 2000). In fact the EP, at its first reading in 1997, adopted a number of stricter amendments that the Commission argued against on grounds of cost effectiveness. However, the Council was in line with the European Parliament on petrol, if not emission and diesel fuel questions. Thus the Council's common position displeased the producer lobby, but only partly won over the environmental groups. Riding on the back of this limited success, the environmental alliance has pushed at the EP and Council for reviews on stricter standards in 2005. As a result, the preceding case shows how the EU institutions created and eliminated a number of policy options and altered the nature of lobbying in Brussels. However, for all the changes in EU agenda-setting and norm creation, national alliances were seen to be of equal importance in reformulation at the EP and Council, and will also have the greatest effect at the period of implementation. Furthermore, the success of the environmentalists may ultimately have had more to do with the fact that the automobile and oil producers were divided over who should absorb the greatest tax burden (Grant 2000).

As the preceding illustrates, environmental groups have suffered a number of problems of organizing at the EU level. While the Commission provided small amounts of funds to facilitate the mobilization of groups such as European Environmental Bureau, Friends of the Earth, and World Wildlife, environmental groups are still underrepresented at the EU level, relative to member state capitals or Washington, D.C. Due to their limited funds, a leading group of environmentalists came together to pool resources and create the "Gang of Seven" (Webster 1998). However, this alliance plus Greenpeace, Climate Network Europe, Birdlife International, and the European Transport and Environment Federation, while having credible political mass, can still only mobilize some thirty people (McCormick 2001). Moreover, due to the varied membership and objectives of the group, ranging from broad ecological issues to the protection of a specific bird, the political focus and influence of the group has often been undermined. This lack of homogeneity in the environmental lobby, both in lobbying style and ideology, makes common ground difficult to identify and has allowed business to "cherry-pick" short life political alliances with the environmental groups in Brussels.

As discussed in the previous section, the hard currency of influence in Brussels is information and expertise. Thus, environmental groups suffered in comparison to business, as they attempted to adapt their domestic public policy models of direct action and media mobilization to the needs of building medium-term issue groups. As Grant (2000) noted, environmental groups are disadvantaged insofar as environmental issues are subject to the vicissitudes of the "issue attention cycle," and popular public support can wane as fast as grow. Effective environmental lobbying in Brussels requires not just developing a legitimate voice and constituency, but also being able to maintain a stable presence over a period of time.

Significantly those environmental groups that have exerted the greatest influence have been groups that have been able to bring specialist knowledge to the table over a long period of time and have created new cross-issue alliances. For example, Climate Network Europe (CNE) has been a central player in the EU climate control debate since the 1980s and has established itself as a core insider (Grant et al. 2000). However, to do this CNE had to form alliances with business lobbies such as the European Association for the Conservation of Energy and European Wind Energy Association, to widen its policy appeal to decision makers.

However, for every successful environmental alliance, we can see numerous occasions where they have been structurally disadvantaged in lobbying the Commission (Grant et al. 2000). For example, most environmental groups are too closely allied with the environmental directorate and the EP, while business plays a complex web across a number of Commission directorates. While having favored access at the Environmental directorate may give them a disproportionate say in the agenda-setting of broad environmental policy, not having the resources to follow the policymaking cycle through various discussions with different directorates and institutions means that they are at a huge informational and policy formulation disadvantage.

For the preceding reasons, we have started to see the new advocacy coalitions and lobbying styles so common in Washington in the 1990s. The result has been that environmental NGOs have started to build informal personal networks and informational reputations in the policy process, which can be utilized on specific lobbying issues at a later date. That said, as happens in Washington, they still have substantial organizational and resource problems to overcome before they can consider themselves playing on a level playing field with business.

Hence, while large businesses continue to gain access by growing the credibility of their identities as European actors, smaller lobbyists and NGOs are compromised by the dominance of national level identities in the representative process. While larger firms enjoy ever-increasing access and influence, smaller interests, despite their centrality to the EU's future economic well-being, find themselves sidelined by uncooperative representatives at the EU level. Although Small and Medium Enterprises (SMEs) and NGOs have made strides in unifying themselves into a single main group, they have some way to go before they present themselves as the coherent and thoroughly European constituency of their larger cousins; other interests, such as consumers and environmentalists, have still to establish a full voice. Only then will they enjoy the bounty of privilege that comes from insider status.

Learning from the Washington Lobbying Experience?

The success of policy forums and a desire to gain access resulted in attempts by large firms to build their European and environmental credentials via the creation of new ad hoc European business alliances, to restructure European federations into large business clubs and to redefine linkages with national political channels and traditional countervailing interests. The resemblance to Washington's "issue networks," the mobilization of grassroots and the Washington business roundtables, is very strong at first glance.

The exact scale of lobbying in Washington is hard to determine, but on environmental issues it is estimated that 3,000 organizations, most of which represent business, have Washington offices. In addition to this there are another 6,000 registered lobbyists and tens of thousands of support staff representing 40,000-plus registered clients, including doctors, senior citizens, foreign governments, religious organizations, and environmental groups and industries affected by environmental issues (Chepesiuk 1994). In addition to these environmental issue groups, the expansion of large firms' direct lobbying of congressional hearings in Washington has been widely catalogued as a consequence of the fast growth of federal government regulation in the 1970s, without a comparable increase in bureaucratic resources (Vogel 1989; Wilson 2003). Thus companies including Monsanto and DuPont have installed specialist lobbying teams on single issues like Superfund reform.

Comparisons of issue networks in Washington and Brussels are complicated by the existence in the United States of a politicized bureaucratic administration and political campaign contributions (Wright 1996). Associations such as the American Petroleum Institute and large companies have Political Action Committees (PACs), whose purpose is to raise and distribute campaign funds for political office. In 1992 elections, PACs totalled more than $172 million, only a small proportion of which came from environmental groups (Chepesiuk 1994). This direct funding gave a large structural advantage to business vis-à-vis environmental and civic groups. However, while environmental NGOs are disadvantaged in funds, the fact that in the United States a number of policy areas have been pushed out of the traditional power centers of federal government and into the intermediary issue networks of the congressional hearings has strengthened the role played by specialist lobbyists (Martin 1991).

Capture is also a risk, as with each changing administration new committee appointments must come up to speed on technical issues and are therefore left exposed to well-prepared and resourced interests (Martin 2000). This risk of capture is greatest where the subsystems have a relatively small number of participants who deal frequently with one another, and when congressmen lack technical expertise (Vogel 1996). Thus, the most successful lobbyists are not necessarily those who paid the highest political contributions, but those who extract the broadest support from the greatest number of actors, and for this reason the U.S. system, like the EU, can be seen to be based upon alliance building, bargaining, and compromises (Sabatier 1998; Wilson 2003).

Learning from U.S. Firms in Brussels

Drawing on the Washington experience, U.S. firms were the most organized and proactive of the early European lobbyists. With seventy U.S. multinationals operating Brussels government affairs offices and the visible political presence of AmCham, U.S. firms demonstrated to their European rivals the importance of direct, regular, and reliable representation at the European Commission. Specifically, AmCham and its EU committee demonstrated the importance of direct firm membership (at the collective European level) and the participation of senior executives with expertise in the policy debate. By adapting the organizational structure around twelve specialized technical committees on issues such as competition, trade, social affairs, and environment, AmCham was able to complement the new European Commission issue-based forums—in fact it was not uncommon for the membership of both committees to be the same. Specifically, this Europeanization of the American business interest was facilitated by the stated policy of AmCham and American firms to hire high-profile European nationals who were known in the Brussels policy community to lead on specific issues. Some saw this as a concerted attempt by AmCham to socialize the more aggressive U.S. lobbying style to the more conciliatory collective business-government relations in Europe (Jacek 1995).

However, while a successful model, the primacy of AmCham in the 1980s was challenged by the arrival of a large number of European firms in the early 1990s. Its position as an agenda-setter was also diminished with the growing importance of the ERT—which had been modelled on the Washington Industrial Round Table. This new club of Europe's senior industrialists soon became the favored big business forum of President Delors and was a major engine in the successful implementation of the Single Market program (Cowles 1995). Consequently, AmCham, while continuing to show selectivity on position papers, created the European American Industrial Committee (EAIC). Today, AmCham, while still solicited directly by the Commission for position papers, has also learned to form ad hoc alliances with the ERT and UNICE.

Learning from the AmCham experience, European federations started to reorganize their structures to incorporate direct firm membership and steering committees. With these changes, American firms' leadership role has strengthened in many sectors; for example, U.S. pharmaceutical and chemical companies were active in restructuring the European Chemical Industry Council (CEFIC), and the European Federation of Pharmaceuticals (EFPIA) (Greenwood 1997). Most significantly, U.S. firms became so integrated within the EU public policy system that they started to participate and chair UNICE subcommittees, as illustrated by Dow's chairmanship of the customs legislation committee and Procter and Gamble's leadership role at both UNICE's Consumer Affairs Committee and AmCham. The acceptance of the U.S. business club/organization model as the most effective means of lobbying the Brussels establishment was to have serious implications for the development of the Commission's business relationship.

Continued Differences in Business-Government Behavior

Despite the apparent desire to emulate many of the lobbying practices of U.S. firms, it was evident that European firms were aware of differences in business culture, state/firm institutional traditions, and codes of political conduct. These differences are illustrated by the discussions focusing on openness and transparency in the U.S. Congress and governance in the EU. For example, the Commission has no register of proposed regulations to solicit public comment, Parliamentary hearings on draft legislation are rare, and virtually no formal advisory bodies exist in the EU public policy system. This informality gives European public policy its vitality and flexibility, allowing for the development of informal relationships, the apportioning of favor and the establishment of trust, encouraging long-run business-government relationships based around committees and expertise. These relationships contrast with the Washington experience, which has tended to be more competitive and to encourage a short-term adversarial culture.

Traditionally, European firms have negotiated industrial and environmental policy in the nation state from a favored and often insiders' position, rather than attempting to block legislation in a confrontational style more typical of the United States. This difference may have some of its origins in the fact that the European nation-state has always been more interventionist and the EU has had to step up its regulatory activity with the creation of the Single Market. In this respect, the Commission has perhaps produced a hybrid of the U.K. and French models, where close but informal relationships have been built between the administration bureaucracies and the business environment (Schmidt 1997).

While the Brussels public debate often appears muted in comparison to Washington and the European policy forums look like small clubs, the reality is that the Commission has taken a number of steps in recent years to encourage openness and transparency. The result is that the Commission should be seen as an informal institution that operates a voluntary code of conduct for interest groups. Whether such openness actually occurs or can survive the increased lobbying in Brussels is sometimes questionable, as we will explore in the following cases. But what is beyond doubt is that large firms believe that they have good access and potential influence at EU institutions at the expense of less well organized and resourced public interests.

However, at the EU level the changing institutional balance, expansion of policy areas, and technical nature of functionaries actually conspired to reduce the chance of bureaucratic capture. Furthermore, with the completion of the Single Market directives, the Commission found itself dealing with new service standard issues and seeing partnerships with wider public interests. Accordingly, with the change in the informational needs and recognition of new European public interests and firms' strong desire to participate in EU policymaking, the Commission was able to restrict/select company access to its 300 committees and 1,200 issue forums. Consequently, by managing access and developing a high technical capability, the Commission believes that it has been able to dictate the terms of the regulatory policy debate.

The technical and regulatory nature of EU legislation also contributes to the low public profile of European business lobbying, as opposed to the more redistributive policy making in the United States, which often leads business and NGOs to attempt to mobilize grassroots public support; a notable exception to this is the recent and very public debate on BSE and British beef sales in Europe (Grant 2000). Generally, however, this low-profile policymaking is reinforced by the democratic deficit in the EU, as bureaucrats within the agencies, forums, and committees are not constrained by public pronouncements at election time, nor do political parties seek funds from interest groups. Thus we are less likely to see the media induced agenda setting and shifting interest group alliances characteristic of the United States.

This may change as the European Parliament attempts to assert itself in specific policy areas via new cooperation procedures with the Commission, and as publicity grows on normative issues such as the environment. Specifically, political accountability of the Commission could increase as the EP asserts itself post-Maastricht and Amsterdam Treaties—as illustrated by the recent censure of European Commissioners Cresson and Marin for fraud and waste (Wallace and Wallace 2000). However, even with the institutional changes in the legislature, it is questionable, whether we will ever see in Europe the same degree of "grassroots" lobbying and "political advertising" that characterizes the U.S. business lobbies.

In the United States, we are seeing new vertical political alliances, as illustrated by lobbying strategies of prescription drug companies against drug-pricing legislation, that include ostensibly grassroots initiatives and the creation of the "Coalition for Equal Access to Medicines." Similarly, "alliance politics" has increased in Europe, as firms have had to develop complex issue identities to access the new elite business forums. Nonetheless, this grassroots activity was motivated less by the desire to develop high-profile press attention than by the need for credibility with individual Commission officials. The democratic deficit and technical nature of governance in Europe continue to ensure that the strategic logic behind ad hoc industrial alliances is different from the United States.

The difference between EU and U.S. political activity can also be attributed to the relative position of the political institutions in Europe and the balance of powers among the branches of government in the United States. In the United States, the Congress acts as an independent actor in the legislative process and firms attempt to influence policy via committees and individual Congressmen. Hence, when the accounting industry and large corporations came together to generate grassroots lobbying for support for the securities legislation reform in Congress, each firm encouraged its employees to write to their Congress members and set up an 800-number that automatically generated letters.

In the EU and member states, the influence of an individual MEP is reduced due to the strength of the party system and the fact that government and parliament can more often be seen as one. At the European level this effect is magnified, as the link between the local interest and MEPs becomes blurred. Hence, MEPs and MPs in the European lobbying context appear less important than U.S. Congressmen and senators. Exceptions can be found on localized issues such as the German shipyard debate on tax subsidies or in high-profile public issue areas involving public interest questions such as genetically modified corn and nuclear energy production. Nevertheless, the potential for party political maneuvering in Strasbourg and the lobbying uncertainty this creates for business means that most budget constrained firms are reluctant to commit resources to lobbying the EP, especially in the current period of recession. Consequently, civil servants at the Commission initiate much of the legislation and are often more knowledgeable than the MEPs on the realities and technical requirements of legislation. This has resulted in a technocratic policy system where firms favor dealing with fellow experts and lawyers. Under these conditions lobbying has become more sophisticated than the cash contributions and aggressive lobbying of American firms.

However, the experience of U.S. and EU firms learning to work together in Brussels has resulted in more cooperation at the international level. Creating new business clubs such as the TABD has shown how far the role of big business in policy formulation has become the accepted norm in Brussels and Washington (Coen and Grant 2001). In learning to collaborate in ad hoc international lobby groups, business has strengthened further its position and voice at intergovernmental treaty negotiations in areas such as climate change and ozone depletion (Levy and Newell 2000).

Conclusions

In understanding European environmental lobbying, it is clear that an elite pluralist environment has evolved where business has a favored position in agenda-setting at the European institutional level. Nevertheless, while the Commission has increasingly become the primary focus of big business lobbying, member-state channels and other EU institutions continue to be of great significance to a well-structured lobbying strategy that follows the policy cycle from agenda-setting through to implementation. Thus, firms must engage with international bodies, European institutions, member state governments, and local authorities on an issue-by-issue basis, and alter their political strategies accordingly.

Accepting the multiple institution approach to environmental lobbying, firms have learned to "mix and match" their political alliances with various environmental and business interests groups to create flexible advocacy coalitions. Firms have learned that favored access to the policy process is about building issue identities and credibility over time, and that the best way to establish reputation is through conciliatory and collective representation via inclusive mutual recognition business-policy forums. Thus, business responsiveness to these new international opportunity structures in environmental politics has created a high level of political convergence in international lobbying strategies.

Clearly, the EU environmental policymaking process has many similarities with the flexible and multichannelled U.S. pluralist model, and European firms have learned from the Washington lobbying experience. What is more, in business-government arrangements it is possible to see issue networks on climate reforms that operate on both sides of the Atlantic and share many lobbying characteristics. In spite of this, while alliance building and regular professional trans-Atlantic business contacts are desirable, European business-government attitudes and practices continue to differ significantly in the state-level management of environmental policy. Specifically, the Commission and European Parliament, in regulating the Single Market program, have sought to consult with a wider variety of economic, societal, and environmental interests via complex forums and formal consultation processes. In this context the European environmental lobbyists have strengthened their potential agenda-setting niche and embedded themselves in the formulation process by "cooperative lobbying strategies" with business to a greater extent than their American counterparts.

References

Aspinwall, M., & Greenwood, J. (1998). Conceptualising collective action in the European Union. An introduction. In J. Greenwood & M. Aspinwall (Eds.), *Collective action in the European Union* (pp. 1–30). London: Routledge.

Broscheid, A., & Coen, D. (2002). Forum politics in Brussels: A game theoretic analysis. *Max Planck Institute for Study of Society, Discussion Series*, Cologne.

Broschied, A., & Coen, D. (2003). Insider and outsider lobbying of the European Commission: An informational model of forum politics. *European Union Politics, 3*(2), 7–32.

Chepesiuk, R. (1994). Who plays it on Capitol Hill and how. *Environmental Health Perspectives, 102*(8). Available at <http://ehp.niehs.nih.gov/docs/1994/102-8/soi.html>.

Coen, D. (1997). The evolution of the large firm as a political actor in the European Union. *Journal of European Public Policy*, 4(1), 91–108.

Coen, D. (1998). The European business interest and the nation state: Large-firm lobbying in the European Union and member state. *Journal of Public Policy*, 18(1), 75–100.

Coen, D., & Dannreuther, C. (2002). When size matters. Europeanisation of large and SME business-government relations. *Politique Europeene*, 7(1), 116–138.

Coen, D., & Grant, W. (2001). Corporate political strategy and global public policy: A case study of the transatlantic business dialogue. *European Business Journal*, 13(1), 37–44.

Coen, D., & Héritier, A. (2000). Business perspectives on German and British regulation: Telcoms, energy and rail. *Business Strategy Review*, 11(4), 29–37.

Cowles, M. G. (1995). Setting the agenda for a New Europe: The ERT and EC 1992. *Journal of Common Market Studies*, 13, 501–520.

Cowles, M. G. (1998). The changing architecture of big business. In J. Greenwood & M. Aspinwall (Eds.), *Collective action in the European Union. Interests and the new politics of associability* (pp. 108–126). London: Routledge.

Cowles, M. G. (2001). The TABD and domestic business-government relations: Challenge and opportunity. In M. G. Cowles, J. Caporaso, & T. Risse (Eds.), *Transforming Europe* (pp. 159–179). Ithaca, NY: Cornell University Press.

Egan, M. (2001). *Constructing a European market*. Oxford: Oxford University Press.

European Commission (2001). European governance: White Paper: Com 2001: 428: Final.

Grant, W. (2000). Pressure groups and British politics. *Contemporary Political Studies*. London: Macmillan.

Grant, W., Matthews, D., & Newell, P. (2000). *The effectiveness of European Union public policy*. London: Macmillan.

Greenwood, J. (2003). *Interest representation in the European Union*, 2nd Ed. Basingstoke: Palgrave.

Greenwood, J., & Aspinwall, M. (Eds.) (1998). *Collective Action in the European Union: Interests and the new politics of associability*. London: Routledge.

Héritier, A. (2001). *Policymaking and diversity in Europe*. Cambridge, UK: Cambridge University Press.

Héritier, A., Kerwer, D., Knill, C., Lehmkuhl, D., Teutsch, M., & Douillet, A. (2001). *Differential Europe—EU impact on national policymaking*. Boulder, CO: Rowman Littlefield Publishers.

Jacek, H. (1995). The American organization of firms. In J. Greenwood (Ed.), *European casebook on business alliances* (pp. 197–207). London: Prentice-Hall.

Jones, R. (2001). *The politics and economics of the European Union*. London: Edward Elgar.

Jordan, A. (2002). *The europeanisation of British environmental policy: A departmental perspective*. London: Palgrave.

Kohler-Koch, B. (1996). Catching up with change. The transformation of governance in the European Union. *Journal of European Public Policy*, 3(3), 359–380.

Levy, D. L., & Newell, P. (2000). Oceans apart? Business responses to the environment in Europe and North America. *Environment*, 42(9), 8–20.

Marks, G., Hooghe, L., & Blank, K. (1996). European integration from the 1980s: State centric versus multi-level governance. *Journal of Common Market Studies*, 34, 341–378.

Martin, C. (1991). *Shifting the burden*. Chicago: University of Chicago Press.

Martin, C. (2000). *Stuck in neutral*. Princeton, NJ: Princeton University Press.

Mazey, S., & Richardson, J. (1993). *Lobbying in the European community*. Oxford: Oxford University Press.

Mazey, S., & Richardson, J. (1996). The logic of organisation: Interest groups. In J. Richardson (Ed.), *European Union power and policy-making* (pp. 200–216). London: Routledge.

Mazey, S., & Richardson, J. (2001). Interest groups and EU policy making: Organizational logic and venue shopping. In J. Richardson (Ed.), *European Union: Power and policy making*, 2nd Ed. (pp. 217–235). London: Routledge.

McCormick, J. (2001). *Environmental policy in the European Union*. London: Palgrave.

Pijenburg, R. (1996). EU-lobbying by ad-hoc coalitions: An exploratory case study. *Journal of European Public Policy*, 5(2), 303–321.

Pollack, M. (1997). Defused interests in Europe. *Journal of European Public Policy*, 4(1), 572–590.

Pollack, M., & Shaffer, G. (2001). *Transatlantic governance in the global economy*. Boulder, CO: Rowman and Littlefield Publishers.

Richardson, J. (1996). Policy-making in the EU: Interests, ideas and garbage cans of primeval soup. In J. Richardson (Ed.), *European Union Power and Policy-Making* (pp. 1–21). London: Routledge.

Richardson, J. (2000). Government, interest groups and policy change. *Political Studies*, 48(1), 1006–1025.

Richardson, J. (2001). Policy-making in the EU: Interests, ideas and garbage cans of primeval soup. In J. Richardson (Ed.), *European Union: Power and policy-making*, 2nd Ed. (pp. 3–24). London: Routledge.

Sabatier, P. (1998). The advocacy coalition framework: Revisions and relevance for Europe. *Journal of European Public Policy*, 5(1), 98–130.

Sbragia, A. (2000). Environmental policy: Economic constraints and external pressures. In H. Wallace & M. Wallace (Eds.), *Policy-making in the European Union* (pp. 293–316). Oxford: Oxford University Press.

Schmidt, V. (1997). European integration and democracy: The differences among member states. *Journal of European Public Policy*, 4(1), 128–145.

Streeck, W., & Schmitter, P. (1991). From national corporatism to transnational pluralism: Organised interests in the single European market. *Politics and Society*, 19(2), 133–164.

Tenbucken, M. (2002). *Corporate lobbying in the European Union: Strategies of multinational companies*. Frankfurt: Peter Lang Publishers.

Vogel, D. (1989). *Fluctuating fortunes: The political power of business in America*. New York: Basic Books.

Wallace, H., & Wallace, W. (2000). *Policy-making in the European Union*, 4th Ed. Oxford: Oxford University Press.

Webster, R. (1998). Coalition formation and collective action: The case of the Environmental G7. In J. Greenwood and M. Aspinwall (Eds.). *Collective action in the European Union* (pp. 178–195). London: Routledge.

Webster, R. (2002). The nature and context of public interest coalitions in the European Union. *Politique Europeene*, 7(1), 138–159.

Wilson, G. (2003). *Business and politics: A comparative introduction*. London: Palgrave.

Wright J. (1996). *Interest groups and Congress: Lobbying, contributions, and influence*. Boston: Allyn & Bacon.

Young, A., & Wallace, H. (2000). *Regulatory politics in the enlarging European Union: Weighing civic and producer interests*. Manchester: Manchester University Press.

IV

The Privatization of Governance: Business and Civil Society

9

The Privatization of Global Environmental Governance: ISO 14000 and the Developing World

Jennifer Clapp

The role of private business actors in global policy making has captured the attention of analysts in recent years.[1] Part of the reason for this interest is the growth in the number of voluntary codes of conduct for firms and private standards-setting bodies, which are gaining recognition and public status among states and intergovernmental organizations (Gereffi, Garcia-Johnson, and Sasser 2001; Nash and Ehrenfeld 1996). This has led to the establishment of what some have labeled as "mixed" regimes of a hybrid nature, whereby both states and private authorities are heavily involved in the creation and maintenance of international principles, norms, rules, and decision-making procedures. In such hybrid regimes, the boundary between public and private spheres is blurred (Haufler 1993, 101). In the global environmental realm, the ISO 14000 series of environmental management standards recently adopted by the International Organization for Standardization (ISO) is illustrative of a hybrid private-public regime, as it was developed and is maintained primarily by industry actors, with some involvement by state representatives. Because of this heavy representation by industry, international standard setting with respect to environmental management systems has become "privatized." These standards are intended to help firms take environmental considerations into account in all aspects of their operations by establishing an environmental management system (EMS) and other operational guidelines. Over 13,000 firms in 75 countries had obtained certification to the ISO 14001 standard by the end of 1999 (Morrison et al. 2000: 2).

As firms in both developed and developing countries adopt these standards, expectations are growing that adherence to the standards will become a de facto condition for conducting business in the global marketplace. Though strictly voluntary for firms, the ISO 14000 standards are extremely important, as their impact goes far beyond private industry practices. States are placing great faith in voluntary industry efforts to help improve environmental quality, and are in effect delegating some authority to them (Agenda 21, chapter 30). The ISO 14000 standards, in particular, are being adopted by standards-setting bodies in some states, either wholesale or in part, as national EMS standards. Moreover, these standards are now recognized by the World Trade Organization (WTO) as legitimate public standards and guidelines. Industry has also been very supportive of the ISO 14000 standards, as it hopes that adherence to them may preempt, or at least soften, present and future state-determined command-and-control type environmental regulations which are much more stringent with respect to performance (Kollman and Prakash 2001, 404).

While industry-based voluntary environmental measures such as the ISO 14000 EMS standards have gained initial encouragement from states and industry-based organizations, debates over them have emerged. The legitimacy of the ISO to establish global norms for environmental behavior is in question, especially since the ISO 14000 standard setting process has been dominated by industry. Moreover, it is not clear that the membership and procedures of the ISO are open and participatory. Some have argued that there must be a proper assessment of these privately set standards, especially where they have gained public recognition, as in the case of ISO 14000 (Krut and Gleckman 1998). There are also debates over whether the standards will in fact lead to environmental improvement. Industry advocates see these new environmental codes of conduct as representing positive rather than negative incentives for good environmental behavior. Critics, however, see voluntary industry environmental standards as a new tool of industry to "greenwash" their operations, while environmentally dubious practices persist, particularly in the developing countries of Asia, Africa, and Latin America (Chatterjee and Finger 1994).

Thus far there has been a great deal of attention paid to the impact of these developments for industrialized countries. But the ISO 14000 standards are also of particular concern for developing countries that do not have as much representation as industrialized countries in organizations such as the ISO. As developing countries increasingly seek to achieve or sustain industrial growth and to break into, or maintain, their share in global export markets for their industrial products, there is growing global concern over the environmental consequences of this rapid growth. For these reasons, it is important to assess the impact of the ISO 14000 standards, both in terms of the legitimacy of the standard-setting process and the level of protection afforded by such standards for developing countries.

In this chapter, I argue that the ISO 14000 standards represent a privatization of global environmental governance, and that this has potentially negative implications for developing countries. Though the standards are widely recognized by states and did involve some state involvement in their establishment, business and industry players exerted direct influence over the standard setting process. The increased involvement of private authority in the setting of environmental standards for industry and their recognition by states and international organizations raise two serious issues for developing countries. The first is that the process of developing these standards has resulted in developing countries losing some of their voice in the development of norms for global cooperation on environmental issues. The second is that these standards may not make much difference in terms of environmental quality in developing countries.

Private Actors in Global Environmental Governance

International relations scholars have noted over the past few decades that nonstate actors have gained in importance regarding their role in supplementing state-based efforts to promote global cooperation (Rosenau 1991). As a result of these developments, there has been growing attention to the theoretical implications of nonstate forms of governance. A great deal has been written in recent years, for example, on the rising presence and impact of "global civil society," NGOs and multilateral institutions on transnational relations broadly, and on the formation of international regimes (Lipschutz 1992; Weiss and Gordenker 1996). Private, market-based actors, such as multinational corporations

(MNCs) and business advocacy associations, while receiving somewhat less attention, are also extremely important actors in this respect and are only now gaining the attention of international relations scholars. These actors influence structures of global governance in several important ways, both indirectly and directly. Their sheer economic weight in global production relations gives them some degree of indirect influence over states (Gilpin 1975; Vernon 1971). And as many critical theorists assert, they are able to exert a diffused but important influence as a result of their dominant role in the formation of broader ideological norms (Cox 1987). They also attempt to exert direct influence through lobbying of state actors involved in the negotiation and maintenance of global rules (Barnet and Muller 1974; Milner 1988). And finally, market-based actors have even more direct influence through the development of "private regimes" or have access to channels for direct participation in the development of hybrid public-private regimes (Cutler, Haufler, and Porter 1999; Stopford and Strange 1994).

Together these direct and indirect channels of influence indicate that private actors indeed play an important role in the formation of global regimes. The involvement of private market-based international authorities in setting international norms is not new, as private institutions have sought, at least since the nineteenth century, to create structures of international governance that would help to facilitate business in a liberal world economy (Murphy 1994). Indeed, scholars from a Gramscian perspective have long highlighted this involvement of business actors as key players in the development of ideological norms that inform the structures of global governance. But since the Second World War, states have taken on a greater role in the formation of formal treaties and regimes, such that an increased direct role for private authorities in recent years appears to be a relatively novel development. This is especially true in the realm of global environmental governance, where initial efforts at global cooperation regarding environmental quality were dominated formally by states. And because regime theory approaches to international cooperation have dominated the study of global environmental governance, the direct and indirect channels of influence of private actors has been downplayed. An approach that leans more toward the Gramscian perspective recognizes that business actors have always had an indirect influence in the formation of regimes, but in a significant shift in the balance of social forces, more direct channels now appear to be opening up for these actors via industry-set standards and codes of conduct.

States have increasingly accepted a more direct role for private actors in the formation of regimes in recent years. While most state-based regimes today have at least some participation from private actors, we are seeing a growing number of regimes in which the balance of influence in decision making is tilted toward private actors. Some of these regimes are in fact initiated by nonstate actors and later endorsed by states as being consistent with state goals. In some cases, states will not only recognize regimes initiated by nonstate actors, but may incorporate them into their own regulatory structures (Haufler 1993, 95). Private actors may actively pursue such endorsements, as a way to embed their own norms into existing global structures. The effect is an increased public significance for privately instigated regimes.

A common explanation for this trend is that in an era of economic globalization, the state has lost some of its control, and thus authority, over certain functions, with the result that market-based actors have begun to fill that void. Strange (1996, 5–10), for example, has argued that private market-based nonstate authorities have gained in significance because the state has lost its effectiveness in areas that the "free market" has never been able to provide adequately, such as security, monetary stability, law enforcement, and public goods. The reason for this decline in effectiveness, she argues, is the rapidly accelerating pace of technological change and the rising costs associated with it, trends that are weakening the state in comparison to nonstate private actors. As the authority of the state is declining in these areas, other forms of governance are replacing it, with states unable, and even unwilling in some cases, to challenge them.

While this sort of argument implies that states have lost some of their ability to govern and that private market-based actors pose a significant challenge to states' authority, the relationship between public and private forms of governance appears to be more complex. Though decreased financial and technical capacity of the state to regulate may be an important factor, the state may delegate authority to nonstate actors, and market-based actors and organizations in particular, not because it is unable to provide the functions that it once held; rather, it may actively choose to do so regardless of its capacity to regulate. Cutler, Haufler, and Porter (1999, 337–338), for example, argue that the state might see private arrangements as being more efficient and more compatible with its preference for a limited role for government in the economy. Such explanations suggest that state and private interests have converged in some areas, particularly with states' renewed adherence to liberal norms in recent years that call for a reduced role for the state (Cutler 1995). In other words, the indirect influence that private actors have in the formation of broader ideological norms may in fact lead the state to open up more direct channels for their influence over governance. Industry may engage in strategic behavior by cultivating an image of itself as being capable of efficient, responsible, and effective self-regulation. In this way it builds alliances with the state in order to solidify its channels of influence. It does this by offering expertise that states often lack, and by forming government-industry "partnerships." By making the argument that its own preferences are likely to be the most efficient and cost effective, industry has been able to project its own interests as the general interest, and hence insert its ideologies and practices into structures of governance.

A clear example of private actors exerting both direct and indirect influence over the formation of global environmental governance is the proliferation of voluntary codes of environmental conduct and private standards for industry. In addition to the ISO 14000 standards, other codes of environmental conduct that have been established in recent years include the Coalition for Environmentally Responsible Economies' (CERES) Principles, the Responsible Care Program of the American Chemistry Council (ACC, formerly the Chemical Manufacturers Association, or CMA), and the International Chamber of Commerce's Business Charter for Sustainable Development. These voluntary environmental codes were established by business advocacy associations in the late 1980s and early 1990s (Nash and Ehrenfeld 1996; United Nations Conference on Trade and Development [UNCTAD] 1996a). Some of these environmental codes were developed in a relatively open process, such as the CERES Principles, which were in fact first suggested by environmental NGOs. Others, however, such as the ICC Business

Charter for Sustainable Development, the Responsible Care Program, and the ISO 14000 series have been more closed, with their content largely determined by industry representatives, though each in principle does allow nonindustry representation.

These voluntary codes operate on the principle that firms should set their own environmental goals and that they should establish an EMS to enable them to work toward meeting those goals; that is, environmental considerations are to be taken into account at all stages of production and marketing. Proponents of these codes argue that they could shift some of the costs of regulation, in particular monitoring of compliance, to the private sector, thus easing the state's burden. But while firms are asked to set their own environmental goals and are asked to commit to preventing pollution, none of the codes mentioned above stipulates that firms must meet specific performance or emissions standards. This is where they differ substantially from state-based regulations, which are usually performance based. This has led to criticisms that the regulatory process has been "captured" by industry interests portraying themselves as environmentally minded (Williams 2001). This capturing of environmental regulations is seen by some to be part of a broader strategy to develop, in Leslie Sklair's words, a "sustainable development historical bloc" (Sklair 2001, 206–209). States appear to have accepted the broader win-win discourse, with companies portrayed as benevolent stewards of the environment pursuing the common interest (Levy 1997). The construction of a specific interest as the general interest, as in the case of ISO 14000, is at the heart of the formation of a historical bloc for environmental management, as it provides business with a claim to moral and intellectual leadership.

While the ISO 14000 series was not the first of these industry-based structures of environmental governance to emerge, it is rapidly gaining wide recognition amongst firms, states, and intergovernmental organizations, eclipsing other voluntary industry environmental initiatives. The standards can be seen as an emergent international regime, as they follow set principles, norms, rules, and decision-making procedures for environmental management systems (Krasner 1983). There are a dozen standards being developed in the ISO 14000 series covering six separate areas, including environmental management systems, environmental

auditing, environmental labeling, environmental performance evaluation, life-cycle assessment, and terms and definitions (ISO 1996a, 2). The
first five standards in the ISO 14000 series, ISO 14001 *Environmental
management systems—specification with guidance for use*, ISO 14004
*Environmental management systems—general guidelines on principles,
systems and supporting techniques*, ISO 14010 *General principles on
environmental auditing*, ISO 14011 *Auditing of environmental management systems*, and ISO 14012 *Qualification criteria for auditors*, were
adopted by the ISO's membership in mid-1996. Firms can obtain certification only for the ISO 14001 standard, whereas the other standards
are to serve as guidance documents (ISO 1996b, 7). Firms in both developed and developing countries are now seeking to learn more about how
to adhere to these new standards, as it is widely seen that these standards will become a condition for firms that wish to compete in the
global marketplace (Denton 1996, 715).

The 14001 standard for environmental management systems, as one
of the first of the series to be adopted and the only one of the standards
for which firms can obtain certification, has received the most attention
thus far. The certification of firms to the 14001 standard is site specific;
that is, each individual facility must apply for certification separately.
The main criteria for obtaining certification are the following (Benchmark Environmental Consulting 1996, 3):

• Each facility must have its own environmental policy statement that
indicates that it is committed to comply with all applicable environmental laws in the jurisdiction in which it is located, and it must be committed to continual improvement and the prevention of pollution.

• The facility must adopt a management system that ensures that it stays
in conformance with its own environmental policy statement.

• The facility must also be audited to ensure that the management system
is indeed implemented. Firms can self-certify or can opt to be certified
by a third-party auditor.

• Suppliers and contractors are to be encouraged to establish their own
environmental management system that conforms to the ISO 14001
standard.

• The certification awarded to firms meeting these criteria must be made available to the public upon request.

Industry has been especially interested in the ISO 14001 environmental management standard because it allows firms to set their own environmental goals and to work out their own way to achieve them. The benefits of obtaining certification from the firm's standpoint are several. The standards are seen as less intrusive than the command and control type of regulation imposed by governments, and there is hope that states may relax environmental regulations or monitoring for those firms that become certified to the standard. Further benefits to certification for firms are that they can use it as a marketing tool, as it enhances the environmental image of the firm to both consumers and other firms that may require that suppliers become certified. It may also enhance firms' image with investors, lead to less expensive insurance premiums, and increase access to capital from lending institutions (Denton 1996). The ISO 14000 standards are also especially attractive to multinational corporations because they are increasingly accepted as global standards, reducing the need to adapt to a number of different national environmental management standards. Also, large firms support EMS standards because they are costly to smaller firms, and constitute barriers to entry (Reinhardt 2000). A further attractive feature of the ISO 14000 standards for firms is that they are less demanding than other voluntary environmental management standards schemes, such as those in Europe, which place more emphasis on environmental performance.

The Roles of Private and Public Actors in the Development of ISO 14000 Standards

The ISO 14000 series of standards can be said to represent a hybrid regime, with significant participation in its establishment and application by both private and public actors. This is seen in the purpose and membership of the ISO itself, the process by which the 14000 series of standards were established, and the way in which they have been embraced by various state and nonstate actors.

The ISO was established in 1946 with the aim of setting technical standards for industry. The principal goal of the organization is to facilitate international trade, technology transfer, and higher efficiency by advancing the global compatibility of goods and standardizing technical specifications. For example, it sets technical standards for a wide variety of products, such as the number of threads on different sizes of bolts, the size and thickness of credit cards, and film speed specifications. ISO standards have been widely adopted by industries around the world, and some governments have incorporated them into their procurement guidelines (Krut and Gleckman 1998).

The ISO is neither strictly a private body nor a public body, but rather is a mix of the two. The ISO claims that it is, in a formal sense, a nongovernmental organization, though UNCTAD refers to it as a private industry-based organization.[2] The membership of the ISO is made up of the standard-setting organizations of 134 countries. The national bodies that constitute the members of the ISO represent a range of governmental bodies and private industry associations. The ISO's membership documentation indicates that 53 percent of its members are government departments, 34 percent are varying shades of mixed private-public bodies, and 13 percent are strictly private organizations. OECD countries account for the vast majority of the ISO's member bodies that are private or have heavy private sector involvement, while non-OECD countries account for the bulk of those members that are government bodies or that have heavy government involvement (ISO 1996c).[3]

The ISO has three categories of members. Full members are national standards setting bodies, correspondent members are organizations that do not have fully developed national standards setting procedures, and subscriber members are countries with small economies who want to be kept informed of ISO developments. The ISO has eighty-five full members, twenty-four correspondent members, and eight subscriber members. Nearly all of the correspondent and subscriber members are from developing countries. Correspondent and subscriber members can attend meetings as observers, but do not participate actively in the technical work of the ISO and they do not have voting rights. Some countries are not represented at the ISO because they lack a national standards-setting body or cannot afford to attend meetings. All industrialized countries have organizations that are members of the ISO, whereas only about half of developing countries are represented in the organization as either full, correspondent, or subscriber members (UNCTAD 1996b, 8, 21–22). The ISO is thus a quasi-private, quasi-public institution that does not have representation from all countries.

Though the ISO is a hybrid organization in terms of its membership, the decision-making process within the organization appears to be dominated by private industry interests, particularly organizations based in industrialized countries. This makes it quite different from intergovernmental organizations in the UN system, where states are generally equally represented. Industry players in hybrid organizations such as this thus have more direct influence in terms of decision making than they do in formal UN-based environmental regimes. The way in which the ISO sets standards is very decentralized. There are some 2,700 technical committees, subcommittees, and working groups associated with the ISO. The participants in these subgroups are made up of delegates from member organizations. In principle, there are representatives from industry, research institutes, government authorities, and consumer bodies involved in these decision-making groups. In practice, most national standards bodies that are members of the ISO do not regularly involve citizens groups or other stakeholders in the development of national positions. Rather, industry is dominant in the process, with some government involvement (Benchmark Environmental Consulting 1996, 12).

The ISO 14000 series of standards for environmental management systems are amongst the most recent to be adopted as official international standards. The impetus for launching discussions on these standards emerged in the early 1990s in response to the preparations for UNCED. The conference organizers had approached the director general of the ISO in 1991 to ask what measures it was taking on the issue of the environment, and whether it was planning to attend UNCED. As the ISO was already discussing whether to venture into the environmental realm, it decided to establish the Strategic Advisory Group on the Environment (SAGE) to discuss what role the organization might play in this arena (ISO 1996a, 2). This group was made up primarily of representatives from national standard-setting bodies, industrial trade associations, private sector firms and consulting firms (UNCTAD 1996b, 29). The general sentiment of this advisory group was that emerging differences in national environmental management systems, such as the British Environmental Management Standard BS 7750 and the European Union's Environmental Management and Audit Scheme (EMAS), could act as potential trade barriers, and that harmonization was desirable (Nash and Ehrenfeld 1996). In 1992, following UNCED, SAGE recommended that an environmental management standard be established by the ISO, with the idea that this would help industry to fulfill the role set out for it in Agenda 21.

The decision to focus on an environmental management standard, as opposed to a performance standard, was an important move for the ISO, as it shifted the organization away from the technical standard setting for which it was known, and toward the setting of "soft" standards with significant public policy relevance. The ISO argued that focusing on standards for management systems, as opposed to performance standards, would provide the flexibility needed for a global standard, as it would only require firms to meet existing environmental regulations, which were different in each country, and would reduce the chances that it would act as a trade barrier. It was further judged by the ISO that a management rather than performance standard would encourage firms to address environmental problems efficiently at their source, rather than at the "end of the pipe" (ISO 1996b, 6–7). The idea for this approach stemmed from the ISO's success with the ISO 9000 quality standards, set up in the 1980s, which were based on improving quality via management systems, rather than by technical specifications.

Negotiations on the ISO 14000 series of environmental management standards began early in 1993 under the auspices of the ISO's newly established technical committee, TC 207, which was set up for that purpose. Canada's standards organization, the Standards Council of Canada, chairs TC 207. About three-quarters of the ISO member bodies that attended the first TC 207 meeting were from OECD countries. Six subcommittees and nineteen working groups of TC 207 were set up to deal with various aspects of environmental issues. In 1996, forty-nine countries had participating member bodies in the activities of TC 207, and seventeen countries had observer members (ISO 1996a, 8). The official delegates to the TC 207 meetings are the representatives of the member bodies, but these participants come from various sectors, depending on whether the member body is a public or a private organization. Industry groups and environmental consultants had a higher representation than other groups in the activities of TC 207 (Benchmark Environmental Consulting 1996, 12).

As an indication of the breakdown of participants in the process from the United States alone, about 400 representatives from U.S. industries (including chemical, petroleum, electronics, and consulting firms) were actively involved, while only about 20 representatives of government and public interest groups were involved (Nash and Ehrenfeld 1996, 37). The drafting committee for the standards was made up mainly of executives from large multinational firms, standards setting bodies, and consulting firms. The chairs of the subcommittees and working groups of TC 207 all came from industrialized countries with half of the working group chairs being employed by major multinational corporations (Krut and Gleckman 1998, 54–55). There have been very few nongovernmental organizations involved in drafting the standards. The World Wide Fund for Nature has criticized the ISO for its failure to involve more developing country standards organizations and environmental NGOs in the ISO 14000 standards setting process. It also complained that the organization suffers from a serious lack of transparency, citing the ISO's adoption in June 1996 of a policy that effectively banned media access to its meetings (Kirwin 1996). This exclusion of the media and lack of NGO participation is very different from UN-sponsored environmental treaty negotiations.

The private-public mix of the ISO 14000 standards goes further than the membership and decision-making processes of the ISO, as states and other international bodies are increasingly recognizing these standards and incorporating them into national standards, governmental regulations, and intergovernmental policies. In this way they are becoming embedded in broader international structures of governance, which in turn increases their legitimacy among states. Asia-Pacific Economic Cooperation (APEC) is promoting ISO 14000 standards as part of its cleaner production strategy in the region (Anon. 1996a, 647). The EU adopted the standards as a European Norm in late 1996, which requires that any national environmental management standards in EU countries that conflict with the ISO 14000 standards must be withdrawn within six months. The United Kingdom withdrew its BS 7750 standard as a result (Anon. 1996b, 863). It is also foreseen that governments will increasingly incorporate them into government procurement policies (Roht-Arriaza, 515–516).

The endorsement of these standards by states may be due, in part, to declining state ability to regulate effectively in this area, as the move appears to be a convenient way to reduce the regulatory burden and cost by shifting the onus of monitoring compliance on to industry itself. However, the fact that states are largely incorporating the standards into their existing regulations, rather than eliminating those regulations altogether, suggests that states have also endorsed these standards in part for ideological reasons. They see the ISO 14000 standards as being consistent with the liberal goal of reducing the role of the state, the retreat from command and control regulation and a move toward self-regulation. States are increasingly "partnering" with industry on such self-regulatory initiatives, and the standards are no doubt seen as a way to shift the burden of implementation to industry. Private actors have clearly picked up on such sentiments in governments in industrialized countries over the past decade, and have taken the lead in establishing the ISO 14000 standards with clear intentions of obtaining at least some regulatory relief in return. The desire for an alliance between states and industry on this issue, then, was mutual.

Intergovernmental organizations are also recognizing the ISO 14000 standards. Though the standards established by the ISO are strictly voluntary, they have gained an important status in the WTO. Under the current Technical Barriers to Trade (TBT) agreement of the WTO, international standards that were either in existence or "imminent" were to be followed by WTO member countries as a way to reduce technical barriers to trade (Benchmark Environmental Consulting 1996, 4–5). Voluntary international standards set by a recognized body such as the ISO are considered by the WTO to be "standards." It does not, however, consider standards set by governments, intergovernmental organizations or UN bodies as "standards." Rather these are considered to be "technical regulations" which the TBT agreement sees as creating potential trade barriers (Krut and Gleckman 1998, 68–70). Because the ISO 14000 standards were being drafted at the time that the WTO agreement was signed in 1994, they were seen as "imminent," and are now recognized as international standards by the WTO. In this way the ISO 14000 standards have become in effect embedded in the existing global trade rules (Finger and Tamiotti 1999).

Under the WTO agreement, states can challenge other states' national EMS standards as a "technical barrier to trade" if they are more stringent than ISO 14000 standards, but states with more stringent standards cannot challenge states with standards weaker than ISO 14000. This endorsement of the ISO 14000 standards by the WTO has thus given them important status in international trade. It has, in effect, created a ceiling for environmental management standards (Benchmark Environmental Consulting 1996, 4–5; Roht-Arriaza 1995). The seal of approval for these standards by the WTO appears to be due mainly to a convergence of interests. The ISO 14000 standards claim to promote freer trade by reducing trade barriers, which matches the liberal trade goals of the WTO.

Implications of ISO 14000 for the Developing World

As many countries in the developing world are seeking to promote export-led industrial development, environmental concerns linked to rapid industrialization have arisen in both developing and industrialized countries. Agenda 21 not only calls on governments to promote environmentally sound technology transfer, but also asks industry to cooperate, particularly through the establishment of EMS. The document also sees an important role for voluntary environmental measures on the part of industry as a way to meet environmental goals such as cleaner production and waste minimization. Given this endorsement of voluntary industry environmental codes, not just by states but also by international organizations as discussed above, it is important to ask whether the ISO 14000 standards will enhance developing countries' ability to meet environmental goals expressed in Agenda 21.

The high degree of influence that private, market-based actors have had on the development of the ISO 14000 standards and the acceptance of those standards among a growing number of states, as well as intergovernmental organizations, have two important implications for developing countries. First, the process by which the standards were developed has not done much to enhance the voice of these countries in global environmental policymaking. The lack of developing country participation in the initial stages of the ISO 14000 standard setting process and poor access to information on the standards in those countries have meant that their concerns were not fully incorporated into the standards. At the same time, the standards may not do much to help improve environmental quality in developing countries.

Developing Countries' Role in Global Environmental Policymaking

Developing countries have a much lower representation in the ISO generally, and they were conspicuously underrepresented in TC 207 in the early stages of setting the ISO 14000 standards in particular. Only two developing countries (Cuba and South Africa) had representatives at the first TC 207 meeting in early 1993. When the ISO 14001 and 14004 standards were voted on and approved as draft international standards in Oslo in June 1995, only six developing countries actively participated in the process up to that time, mainly from the category of rapidly industrializing developing countries. Thus while 92 percent of industrialized countries were present at the Oslo meeting and voted on the standards, only 16 percent of developing countries were present and voting (Krut and Gleckman 1998, 42). In the negotiation of UN-based environmental treaties, developing countries have generally pushed for phase-in periods and economic assistance to enable them to meet the treaties' demands without incurring significant costs which may hinder economic growth. Though they may not have pushed for stricter EMS standards if they had been more active in the earlier stages, representatives from developing countries that were present at the Oslo meeting expressed their disappointment at the lack of attention in the discussions to issues of technology transfer, a phase-in period for developing countries, and commitment to equal representation from developing countries in the standard setting process (UNCTAD 1996b, 22–38).

One of the key factors behind the lack of participation of developing country standards bodies and other representatives is that the cost of attending technical committee and plenary ISO meetings is borne by each individual participant. These costs can be substantial, particularly for representatives from developing countries. At the June 1995 Oslo meeting, there was an increased number of developing country representatives which had joined the TC 207, mainly as a result of financial assistance from the Netherlands and Finland which was channeled through the DEVCO assistance program organized by the ISO. This program is intended to promote environmental management and ISO 14000 in the developing world by raising funds from donors to support developing country representation at TC 207 meetings. Some twenty-two representatives from the developing world were sponsored under this program in 1995, and twenty-three in 1996. The program only funds two representatives per country, generally one from a standards setting organization, if it exists, and one from a government environmental agency or environmental NGO. Thus developing countries had much smaller delegations than those of industrialized countries, making participation in the twenty-five working groups and subcommittees of TC 207 particularly difficult. Through this sponsorship of representatives from developing countries, the ISO was attempting to avoid the mistakes made by the ISO 9000, as developing countries were largely absent from the process of developing those standards for nearly five years.[4] But because developing country participation in the ISO 14000 process only picked up in 1995, at the point of voting for the standards already developed, it is not clear that this goal was met.

In addition to the lack of developing country participation in the actual negotiations of the standards, access to information on the ISO 14000 series has been highly variable in developing countries. Information has been channeled back to standards organizations and firms in some cases, while in other cases it has not. The ISO has acknowledged that lack of access to information was a potential problem, and has helped to fund awareness raising seminars in some developing countries, such as Indonesia, the Philippines, and China. Firms in a number of industrializing developing countries are particularly keen to learn more about ISO 14000 as they see it as potentially a necessary condition for doing business on global markets, especially in environmentally sensitive sectors. It is also seen by these firms as an important marketing tool for selling their exports abroad in the United States and Europe (Anon. 1999, 652). This has prompted some standards bodies in developing countries, such as Thailand, Singapore, and Mexico to introduce pilot projects to help local firms to become more aware of the steps involved in ISO 14000 certification (Anonymous 1996c, 647).

Meeting Agenda 21's Environmental Goals for Industry

It is not clear that the ISO 14000 series of EMS standards will actually improve developing countries' ability to meet environmental goals for industry, as expressed in Agenda 21. One of the main criticisms made with respect to the ISO 14000 series, is that it is based on environmental management and not on environmental performance (Porter 1995, 43–44). Indeed, as Levy has shown in his study of TRI data, environmental practices (such as EMS) and environmental performance are quite unrelated (Levy 1995, 45–46). The ISO 14002 standard does not call for any specific reductions in hazardous waste generation and firms are not required to report emissions levels. Nor are there any stipulations in the standards calling for the transfer of cleaner technologies to developing countries. Firms must only ensure that management systems are dedicated to meet existing environmental laws in the country in which they are operating, and that there is commitment to "continual improvement" and "prevention of pollution." The fact that the standards only require firms to comply with existing laws means that the ISO 14000 standards may be very costly to implement, but may not actually make much difference in terms of environmental quality, especially in developing countries. Indeed, firms that already have a strong EMS in place have found the ISO 14000 standards to be weak at best (Thayer 1996, 11). Moreover, the standards do not even mention existing international environmental treaties as being a concern for firms.

Firms set their own environmental policy and goals under the standard, and are only judged against their own management system to address those self-set goals when they are audited. Consequently, they can set very low goals and still become certified to the ISO 14001 standard, provided that they comply with environmental regulations in the state in which they are operating. This applies to both developed and developing countries, though firms in the latter category may be more tempted to set lower goals because of weaker government regulations. At one information seminar in Thailand, for example, a Western consultant advised Asian firms not to set their goals too high when formulating their own EMS, and to aim only for compliance with existing national environmental regulations (Anon. 1995a, 841). Indeed, because the 14001 standard is not performance based, firms in developing countries, either locally based or MNC subsidiaries, have little incentive to go beyond meeting the existing environmental laws. While developing countries would likely not support the elimination of the differences in environmental regulations between industrialized and developing countries, they have called for the transfer of cleaner technologies.

The 14001 standard provides little incentive for firms to adopt cleaner technologies, and MNCs are not required to transfer such technologies to their subsidiaries in developing countries. There was, in fact, debate at the drafting stage of the standard over whether it should require firms to adopt the best available technologies in order to gain certification. In the end, this requirement was not incorporated, and the draft standard only encouraged firms to consider implementing the best available technology where it was "appropriate and economically viable" (cited in Roht-Arriaza 1995, 506). Joseph Cascio, a key proponent of the standards who was involved in the drafting process, has since remarked that the goal of the standards is not to get every country to use the same best technologies, reasoning that, "If all countries had to use the best available technology, what was supposed to be a management standard would turn into a performance standard, making the standard unworkable and unattainable for many developing countries" (cited in Anon. 1995b, 21). If the standards included more incentive to transfer clean production technologies to developing countries, this concern would be alleviated somewhat.

With the standards structured as they are, there is no clear sign that government environmental standards and regulations in general will be strengthened in developing countries as a result of firms' adherence to the ISO 14000 series. Indeed, existing environmental laws could be watered down in developing countries if the ISO 14000 standards become widely used by firms. Cascio has argued that the standards may lead some countries to discover that they have more laws on their books than they can ever enforce, given their resources. He sees this as providing the impetus for some developing country governments to redraft their environmental laws so that they can meet existing resources and capabilities (Cascio 1996). While this may increase the ability of firms to comply with legal requirements, it could weaken the existing regulatory framework in those countries. An additional reason why the regulatory framework in developing countries may not improve is that with the legitimacy given to the standards in the WTO TBT agreement, it is almost certain that developing countries will not establish national EMS standards more stringent than the ISO standards, as they could be challenged as trade barriers (Krut and Gleckman 1998, 68–70).

MNCs, as well as local firms in developing countries, may also try to use ISO 14000 certification to obtain regulatory relief, which firms are already trying to do in the US and other industrialized countries. Firms in a number of countries are pressing their governments for less stringent monitoring and enforcement of environmental regulations for companies that are certified to the ISO 14001 standard (Mullin and Sissell 1996, 38). Following requests from firms to relax regulatory requirements for firms certified to the ISO 14001 standard, governments in North and South alike, including in the United States, Argentina, South Korea, and Mexico, are responding with a variety of measures that take ISO 14001 into account in the monitoring and enforcement of regulations (Finger and Tamiotti 1999; Speer 1997). These developments may prompt firms in other developing countries, both MNCs and locally based, to push for the same. This would have the effect of weakening the regulatory framework in developing countries.

While the benefits that have been advertised may not be forthcoming, the costs of adherence (and nonadherence) to ISO 14000 standards are expected to be high for developing countries. One of the key costs of not obtaining certification to the ISO 14001 standard is that products marketed abroad by firms from the developing world could be subject to trade discrimination from ISO 14001 certified firms in industrialized countries that require that their suppliers also become certified (Morrison et al. 2000). This prospect is worrying for many firms in newly industrializing developing countries that are seeking to expand their export markets by supplying firms in industrialized countries with parts and products. One firm in Hong Kong, for example, was dropped as a supplier for a British firm because it did not conform to the ISO 14001 standard (Anon 1996d, 890). Thus, there is real concern that certification to the ISO 14001 standard may become a de facto business condition for operating in global markets, as is now the case with the ISO 9000 standard. This is why many developing countries, particularly newly industrializing countries, are promoting the standards among their exporting firms.

If there are costs to noncertification, the costs of seeking ISO 14000 certification can also be very high, both for developing country governments and firms. A significant portion of the cost of certification is the fee charged by the auditor for consultation and development of an EMS, including the first audit. Other costs are fees for audits on a periodic basis to ensure that the firm maintains conformance to the EMS, as well as capital costs incurred in doing so. These costs can vary widely, depending on the size and type of firm and on rates charged by consultants for their services. Estimates for certification fees alone in industrialized countries range from U.S.$ 100,000 to $1 million for large firms, to U.S.$ 10,000 to $50,000 for small- and medium-sized enterprises (Cohen 1998). These costs for firms in developing countries may be significantly less if local auditing and certifying services are in place. If local auditing services do not exist, there will be costs in setting them up. Because most of the standards setting bodies in the developing world, where they exist, are public institutions, this is a cost that governments will likely bear. And even if auditing services are in place, it is not clear that firms certified by local auditors will gain the same recognition in international markets as they would if they certify with internationally recognized, and more expensive, auditors. Meeting such costs will be especially hard on small and medium-sized enterprises, which account for the bulk of firms in most developing countries.

Conclusions

The ISO 14000 standards are a hybrid private-public regime that has gained important public policy status as states and intergovernmental organizations have begun to endorse and incorporate it into their own regulatory structures. Though it has public significance, the formation of the regime was in large part "private" because of the heavy direct influence of industry actors at the standard setting stage. It is not clear that the ISO itself realizes its own weight in this area. The organization still sees itself primarily as a scientific and technological organization that is not so concerned with the development of international environmental law. Ironically, because these standards are increasingly being recognized by public intergovernmental bodies and national governments, it is precisely in the area of international environmental governance that the ISO 14000 standards are having a profound impact. In effect, private actors have "captured" the regulatory process for environmental management system standards, and they have embedded them within international structures for the regulation of trade. At the same time, states, as well as organizations like the WTO, have embraced these standards because they fit their liberal economic ideals, such that the capturing process is two-way and mutually beneficial.

The direct involvement of industry players in the development of these standards is illustrative of a new trend for business influence in global regime formation that consolidates the more diffused and indirect influence that such actors have traditionally held. In this way, the ISO 14000 process can be seen as one example of the way in which industry has contributed to the formation of the broader "sustainable development historic bloc" referred to by Sklair (2001). It appears as though states are embracing these privately set standards as public policy, in part because they fit well with the prevailing liberal ideology that calls for a reduced regulatory role for the state. Private actors themselves have played a role in the development of these norms through their alliance-building with states. The standards are also seen by states and industry to be a way to reduce trade barriers in an era of global free trade. It is not clear, however, that states have lost a significant amount of their ability to regulate in the environmental realm, as rather than replacing state-based environmental regulations with the ISO 14000 standards, states are attempting to rewrite regulations to take these standards into account.

The trend toward increased private sector participation in the development of certain elements of global environmental governance has other important implications as well, especially for developing countries.

The fact that much of global environmental governance in the past twenty years has been based on formal state-based regimes such as international treaties, which themselves are based on one-state, one-vote structures, has given developing countries at least some voice in international environmental governance matters. Generally, they have won concessions such as economic assistance and phase-in periods to enable them to meet the provisions of these state-based treaties. But with the growing influence of private economic actors, developing countries may be losing some of their voice in this realm, as it is in the industrialized countries where much of this type of governance is initially developed. The ISO 14000 standards did not have equal representation from all countries, and they make no reference to phase-in periods, economic assistance or technological transfer. At the same time, however, developing countries are finding that they are profoundly affected by these new structures of environmental governance, particularly when public international bodies and organizations, such as the WTO, embrace them. Moreover, it is not clear that the ISO 14000 series of environmental management standards will, in practice, help to meet environmental goals for industry as established in Agenda 21. The ISO 14000 series is rapidly becoming accepted as *the* international environmental "standard," even though it is now widely seen by critics, as well as some industry experts, as being weak in terms of its requirements.

Notes

1. This is a substantially revised and updated version of Jennifer Clapp (1998) "The Privatization of Global Environmental Governance: ISO 14000 and the Developing World," from *Global Governance*, 4 (3). Portions of that article are used with permission of Lynne Rienner Publishers.

2. Interview with Klaus Gunter Lingner, ISO Official, July 9, 1996, Geneva.

3. There are some exceptions, for example Japan's member body is a government department, while Argentina's is a private organization.

4. Interview with Anwar El Tawil, ISO Developing Countries Program, Geneva, July 9, 1996.

References

Anonymous (1995a). Foreign firms expected to lead the way on ISO 14000 certification in Thailand. *International Environment Reporter, 18*(27), 841.

Anonymous (1995b). ISO 14000 Standards tailored to meet the needs of all countries. *Quality Progress, 28*(6), 21.

Anonymous (1996a). APEC ministers agree to push for "dramatic progress" on ocean cleanup. *International Environment Reporter, 19*(15), 646–647.

Anonymous (1996b). ISO 14001, 14004 standards finalized; UK to withdraw its measure in March. *International Environment Reporter, 19*(20), 863–864.

Anonymous (1996c). More work needed in ISO 14000 process on certification, says committee member. *International Environment Reporter, 19*(15), 647.

Anonymous (1996d). Hong Kong firms indicate ISO 14000 will affect suppliers, contractors. *International Environment Reporter, 19*(20), 890.

Anonymous (1999). Developing countries said to be increasingly viewing environment rules as protectionist. *International Environment Reporter, 22*(16), 652.

Barnet, R., & Muller, R. (1974). *Global reach: The power of the multinational corporations*. New York: Simon and Schuster.

Begley, R. (1996). ISO 14000: A step toward industry self-regulation. *Environmental Science and Technology News, 30*(7), 298A–302A.

Benchmark Environmental Consulting (1996). *ISO 14001: An uncommon perspective*. Portland, ME: Benchmark.

Cascio, J. (1996). The increasing importance of international standards to the U.S. industry community and the impact of ISO 14000. *EM*, November, 16–23.

Chatterjee, P., & Finger, M. (1994). *The earth brokers*. New York: Routledge.

Cohen, J. (1998). More than 4,000 companies certified under ISO 14001: Japan, Europe lead way. *International Environment Reporter, 21*(13), 650.

Cox, R. (1987). *Production, power, and world order*. New York: Columbia University Press.

Cutler, A. C. (1995). Global capitalism and liberal myths: Dispute settlement in private international trade relations. *Millennium, 24*(3), 377–397.

Cutler, A. C., Haufler, V., & Porter, T. (Eds.) (1999). *Private authority and international affairs*. Albany, NY: SUNY Press.

Denton, C. (1996). Environmental management systems: ISO Standard 14000. *International Environment Reporter, 19*(16), 715–717.

Finger, M., & Tamiotti, L. (1999). The emerging linkage between the WTO and the ISO: Implications for developing countries. *IDS Bulletin, Globalisation and the Governance of the Environment, 30*(3), 8–16.

Gereffi, G., Garcia-Johnson, R., & Sasser, E. (2001). The NGO-Industrial complex. *Foreign Policy, 125*, 56–65.

Gill, S. (Ed.) (1993). *Gramsci, historical materialism and international relations.* Cambridge, UK: Cambridge University Press.

Gilpin, R. (1975). *U.S. power and the multinational corporation.* New York: Basic Books.

Haufler, V. (1993). Crossing the boundary between public and private: International regimes and non-state actors. In V. Rittberger (Ed.), *Regime theory and international relations* (pp. 94–111). Oxford: Oxford University Press.

ISO (1996a). *The ISO 14000 Environment.* (March).

ISO (1996b). 'Environment: ISO 14000 Standards Focus on the User.' *ISO Bulletin,* 27(2), 6–7.

ISO (1996c). *ISO members.* Geneva: ISO.

Kirwin, J. (1996). WWF calls on ISO to clear up confusion surrounding extension of ISO 14001 to forestry. *International Environment Reporter,* 19(19), 811–812.

Kollman, K., & Prakash, A. (2001). Green by choice? Cross national variations in firms' responses to EMS-based environmental regimes. *World Politics,* 53(3), 399–430.

Krasner, S. (Ed.) (1983). *International regimes.* Ithaca, NY: Cornell University Press.

Krut, R., & Gleckman, H. (1998). *ISO 14001: A missed opportunity for global sustainable industrial development.* London: Earthscan.

Levy, D. L. (1995). The environmental practices and performance of transnational corporations. *Transnational Corporations,* 4(1), 44–67.

Levy, D. L. (1997). Environmental management as political sustainability. *Organization and Environment,* 10(2), 126–147.

Lipschutz, R. (1992). Reconstructing world politics: The emergence of global civil society. *Millennium,* 21(3), 389–420.

Milner, H. (1988). *Resisting protectionism: Global industries and the politics of international trade.* Princeton, NJ: Princeton University Press.

Morrison, J., Cushing, K. K., Day, Z., & Speir, J. (2000). Managing a better environment: Opportunities and obstacles for ISO 14001 in public policy and commerce. *Pacific Institute for Studies in Development, Environment and Security, Occasional Paper, Executive Summary.* Oakland, California.

Mullin, R., & Sissell, K. (1996). Merging business and environment. *Chemical Week,* 158(38), 52–53.

Murphy, C. (1994). *International organization and industrial change.* Cambridge, UK: Cambridge University Press.

Nash, J., & Ehrenfeld, J. (1996). Code green. *Environment,* 37(1), 16–20, 36–45.

Porter, G. (1995). Little effect on environmental performance. *Environmental Forum,* 12(6), 43–44.

Roht-Arriaza, N. (1995). Shifting the point of regulation: The international organization for standardization and global law-making on trade and the environment. *Ecology Law Quarterly, 22*(3), 479–539.

Reinhardt, F. L. (2000). *Down to earth: Applying business principles to environmental management.* Boston: Harvard Business School Press.

Rosenau, J. (1991). *Turbulence in world politics.* Princeton, NJ: Princeton University Press.

Sklair, L. (2001). *The transnational capitalist class.* Oxford: Blackwell.

Speer, L. (1997). From command-and-control to self-regulation: The role of environmental management systems. *International Environment Reporter, 20*(5), 227–229.

Stopford, J., & Strange, S. (1994). *Rival states, rival firms.* Cambridge, UK: Cambridge University Press.

Strange, S. (1996). *The retreat of the state.* Cambridge, UK: Cambridge University Press.

Thayer, A. (1996). Chemical companies take wait and see stance toward ISO 14000 standards. *Chemical and Engineering News, 74*(14), 11–15.

United Nations Conference on Trade and Development (UNCTAD) (1996a). *Self-regulation of environmental management.* Geneva: United Nations.

United Nations Conference on Trade and Development (UNCTAD) (1996b). *ISO 14001: International environmental management systems standards: Five key questions for developing country officials* (Draft report). Geneva: United Nations.

United Nations. (1992). *Agenda 21.* New York: United Nations.

Vernon, R. (1971). *Sovereignty at bay.* New York: Basic Books.

Weiss, T., & Gordenker, L. (1996). *NGOs, the UN and global governance.* Boulder, CO: Lynne Rienner.

Williams, M. (2001). In search of global standards: The political economy of trade and the environment. In D. Stevis & V. Assetto (Eds.), *The international political economy of the environment: Critical perspectives* (pp. 39–61). Boulder, CO: Lynne Rienner.

10

Privatizing Governance, Practicing Triage: Securitization of Insurance Risks and the Politics of Global Warming[1]

Sverker C. Jagers, Matthew Paterson, and Johannes Stripple

Security is a commodity bought like any other: and as its rate of tariff falls in proportion not with the misery of the buyer but with the magnitude of the amount he [sic] insures, insurance proves itself a new privilege for the rich and a cruel irony for the poor.

—Proudhon, as quoted in Ewald (1991, 206).

Introduction

Gramsci's concept of hegemony suggests that dominant groups must negotiate, amongst themselves and in relation with subordinate social forces, to maintain the consensual basis for a stable social formation. Successful construction of hegemony thus requires the "ability to project a conception of the general interest" (Levy and Newell, chapter 3). In the context of this focus on negotiation and coalition-building, negotiation of a specific environmental treaty or the adoption of environmental management practices are examples of a process during which successful accommodations to social pressure preserves the broader hegemonic position of business. Levy and Newell (chapter 3) identify three sites where hegemony is reproduced. These are the economic system, organizational capacity, and the discursive structure. If we examine the climate case more specifically, we can see these different elements of hegemony at work. On the economic level, companies apply strategies and technologies that protect their existing market positions, and invest some resources in environmentally preferable products and processes. For example, in 1997–1998, some of the more influential oil companies (notably Shell International and British Petroleum) "redirected" their business from being oil-focused to become energy companies, investing some of their economic turnover on alternative energy research (Skjaerseth and Skodvin 2001). On the organizational level, companies often form issue-specific associations to lobby, mobilize resources, and coordinate strategy, and they frequently attempt to forge new alliances in order to broaden the historical bloc. The fossil fuel industry has attempted to forge such alliances in organisations such as the Pew Center or the Business Council for a Sustainable Energy Future.

Finally, on the discursive level, companies attempt to challenge the scientific and economic basis for regulation, and use public relations to portray themselves and their products as "green." They adopt a language of sustainability, stewardship, and corporate citizenship. The energy corporations can serve as a good example of this strategy, where they, for example, sell "green diesel" and invest money into alternative energy research. By adjusting their rhetoric, based on minor adjustments in their business practices, while at the same time collaborating with potential pressure groups, business attempts to become impregnable against critique and further scrutiny. Such green marketing exercises clearly create possibilities for campaigns against companies highlighting their "greenwash" nature. In general, however, their effect is to deflect criticism and reproduce corporate hegemony.

But at the same time, hegemony is always contested, as there are differing accounts of the general interest, and varying attempts to construct coalitions around such accounts. The "historical bloc" and the dominant actors within it are constantly exposed to challenges from other social groups. For the bloc not to be weakened, attempts are made to bring moderate elements into a ruling coalition during periods of "crisis." By doing so, the possibility of contesting hegemony is spread among a wider range of subordinate groups. Crises, such as that posed by a phenomenon like climate change, create particularly useful openings for actors wishing to advance such alternatives. Such crises can arise from changes in markets, regulations, technology, as well as relative power positions and ideology, and while it lasts, the historical bloc is both fragile and open to challenge. This temporary vulnerability is often used by less influential actors with fewer material resources to initiate, what might be considered "counterhegemonic projects."

Our chapter discusses one such attempt within climate change politics. In the early 1990s, some elements within the environmental movement, notably Jeremy Leggett, then one of the main climate campaigners for Greenpeace International, made strong approaches to insurance companies to try to persuade them of the risks of climate change. In Gramscian terms, Greenpeace was engaged in a (mini) counterhegemonic project, to construct a coalition of forces, embracing elements of capital alongside civil society organizations and movements, to challenge coal and oil interests in climate change policy debates. While Leggett never himself articulated it as such, this represented an attempt to prise open the historic bloc comprising the dominant corporate response to climate change, by splitting apart an element within that bloc whose interests are threatened directly by climate change itself. Later, from 1995 onwards, the project was taken up by UNEP (United Nations Environment Programme), and remains an UNEP-led project (Leggett left Greenpeace in 1997 to found Solar Century and the Greenpeace insurance project was a "Leggett" operation).

Obviously, this can only be interpreted as counterhegemonic in a very narrow sense. It is clearly not the case that insurers are a possible ally in a broad counterhegemony against neoliberalism, since, as Adam Harmes (1998) shows, insurers have been part of a block of institutional investors helping to reproduce neoliberalism through a variety of coercive and consensual means.[2] In addition, Leggett and UNEP never conceptualized the project in these broad terms.

The Politics of Counterhegemony

In the context of this general theoretical and empirical terrain, we try to analyze the politics of this attempt at counterhegemony in this chapter. The language of hegemony is useful here, since it involves an attempt by social movements and international institutions to forge alliances with elements of capital. Such an alliance materially challenges the power of other elements of capital in bargaining with states, organizationally challenges the established actors on the climate arena by broadening the issue-specific associations, and discursively challenges the dominant notions that understand action to mitigate climate change to be an economic constraint rather than an opportunity.

But this attempt to construct such a counterhegemonic project remains a case of failure. What we show in this chapter is that in the insurance-climate change case, the emergence of new forms of privatized governance has enabled insurers to limit their need to lobby governments to reduce greenhouse gas emissions (GHGs)[3] and mitigate climate change. This is the case despite the fact that several influential insurers have formally signed up to the UNEP insurance initiative, which advocates such cuts, and some also have started to work to invest proactively in renewable energy technologies. The ability of insurers to use new developments of risk-transfer mechanisms has meant that many of them now think they can adapt to a warming world even if the risks of large-scale weather catastrophes increase in such a world. This is the case, even if over the medium-term, the likelihood of such risks occurring becomes nonactuarial. Thus, the counterhegemonic potential of a Greenpeace-UNEP-Insurance coalition has diminished considerably during the last few years.

This is not to say that such strategies of insurers should be understood as simply economic actions by rational firms. We argue they should also be regarded as fundamental practices of governance. This is because they regulate the behaviour of others normatively by structuring incentives to locate businesses and homes in particular places, by disciplining the owners of existing and future buildings about their ability to withstand large-scale weather events. Most fundamentally they construct, as all insurance does, some as subjects and others as nonsubjects. This is thus a privatised form of governance, whereby insurance companies arrogate to themselves the decisions about who is to be made subject to large-scale climate related risks.

It is by no means the case, then, that the activities of the insurance industry are politically irrelevant. It only means that the space for possible counterhegemonic projects is limited by the emergence of privatized governance, in that the ability to build political coalitions is narrowed when the elements of capital with which social movements actors might want to construct such coalitions have exit options which states underregulate. The conventional account of the construction of hegemonic projects assumes that there are outcomes, understood in terms of state action, which are the subjects of contestation. Here, where firms have other options than relying on state action, they cease to have to engage in coalition-building to realize their interests. Clearly, in the grander scheme of things, this reflects a large-scale hegemonic success of neoliberalism, which confers rights and capacities to govern on such actors.

Insurers' Concerns about Global Warming

It is worth spelling out briefly the link between emissions of carbon dioxide (CO_2) and extreme weather events and, hence the nature of insurers' concerns about global warming. For even though the number of companies that have been able to translate this concern into action to mitigate climate change has been small, many have been concerned about the impacts this change has for the insurance business (Leggett 1996; Paterson 2001; Stripple 2000; Stripple, Chong, and Wiman 1997). The greenhouse theory holds that a large number of human activities, such as the burning of fossil fuels, increase atmospheric concentrations of GHGs that, in turn, affect the escape of heat from the Earth. The Intergovernmental Panel on Climate Change (IPCC)[4] predicted in its Third Assessment Report (TAR) that the changes in GHG concentrations will lead to a 1.4 to 5.8°C rise (compared to 1990 levels) in global mean surface temperature by 2100 (IPCC WG I 2001; Kerr 2000). Warming is predicted to lead to regional and global changes in climate and climate-related parameters such as temperature, precipitation, soil moisture, and sea level (15–95-cm rise by 2100) (Intergovernmental Panel on Climate Change Working Group II [IPCC WG II] 1995, 3).

The main worry of insurers has been that many of the projections of global warming have suggested that it would involve increases in the incidence, severity, and location (specifically, moving away from the Equator) of catastrophic weather events. Floods, but especially windstorms, are of particular concern for insurers. Alongside scientific developments, this concern was stimulated by an extreme and rapid increase in the payouts to such disasters in the late 1980s. Insurers noted that payouts on such catastrophes had noticeably increased from the mid-1980s onwards (Kron 2000). Economic damages caused by natural disasters had exceeded $20 billion in only two years prior to 1988; after that date, they have been lower than $20 billion in only one year—1997 (Munich Reinsurance 1998; Oberthur and Ott 1999, 74).[5] This increase is accounted for primarily by damages from wind storms and flooding. In the 1980s, the damage from wind storms was $3.4 billion, while in the first three years of the 1990s alone it was $20.2 billion. The largest wind storm was Hurricane Andrew, which caused $16–17 billion of damages, and several small insurers in Florida to go bankrupt (Dlugolecki 1996, 69).[6]

For some, the potential threats to the interests and survival of insurers, and by extension, to large parts of the international financial system, from these catastrophes was great. The losses from Hurricane Andrew raised the possibility that one storm alone could cause such damages, if it hit a major city such as Miami, Washington, D.C., or even possibly New York directly. In particular, the world's two largest reinsurers, Swiss Re and Munich Re, as well as some Lloyd's syndicates, claimed as early as 1992 that climate change could bankrupt the global insurance industry (Leggett n.d., 26–30; Schmidheiny 1992, 64–66; Schmidheiny and Zorraquin 1996, 121–122). That it was these organizations that led the claims about global warming is perhaps unsurprising. Haufler (1997b, 88) regards Munich Re and Swiss Re as among the most conservative of reinsurers in assessing which risks are insurable, while Lloyd's was undergoing a huge general crisis because of massive losses involving potential bankruptcy of many Lloyd's "Names" (Strange 1996, 131).

At the heart of insurers' worries is that the basis on which they underwrite risks of flooding or storm damage is no longer adequate. Insurance relies fundamentally on being able to assess these risks on an actuarial or probabilistic basis. This enables them to set premiums that will cover payouts. If global warming is changing the incidence and severity of such storms, and especially if it is doing so in a nonlinear fashion, then the data about storms, floods, etc., on which companies have set premiums, are no longer adequate. Worse, setting premiums on an actuarial basis is perhaps impossible given the constant change implied by the process of global warming.

Insurers, Global Warming, and Greenpeace

As highlighted by Levy and Newell in chapter 3, hegemony is never completely stable, and is subject to continuous reproduction and renegotiation. Central for any counterhegemonic project is the ability to open up a space from where alternative conceptions of the general interest can be constructed. On the climate issue, Greenpeace tried very purposefully to open up such a space by exploring a tension between the insurance industry and the fossil fuel industry.

Immediately after the Earth Summit in 1992, Jeremy Leggett of Greenpeace International embarked on a project to persuade insurers of the need for them to become involved in climate change politics.[7] His interest in this was stimulated by an original idea by his colleague at Greenpeace International, Paul Hohnen (Leggett 1999, 100). Leggett and Hohnen had initial ideas about this during 1991 and early 1992 while reading reports of catastrophic losses, in particular to Lloyd's syndicates, due to flooding and windstorms. Leggett reports that an article in *Time* magazine on 8 July 1991 "showed clearly how catastrophes involving extreme climatic events were a major part of the problem emerging in the three-hundred-year-old institution" (1999, 47).

Leggett is worth quoting at length on his (reconstruction of) his original impetus. He writes that during a break from work after the Earth Summit:

I felt sure that a new strategy was needed. I was beginning to feel that governments alone would not and could not deliver the goods . . . I and my Greenpeace colleagues had been frustrated for some time that the Global Climate Coalition and their collaborators had been able, unopposed, to present themselves as the voice of business. We knew they were not. They were the voice of a certain category of business: fossil fuels, automobiles, and to a degree, chemicals. They were not the voice of the fishing industry, the skiing industry, the medical profession— all the sectors that stood to lose in a world making no effort to reduce the enhanced greenhouse risk. They were certainly not the voice of the financial sector, or most especially of the insurance industry. (Leggett 1999, 100)

His construction of the problem is clearly couched in terms of the necessity for social movement organizations to engage with business. This reflects his understanding of the power of business in global environmental politics. But it also reflects an understanding that interests of different sectors of business (or corporations within one sector—such as the variety of insurance companies) are not necessarily always homogenous, and sectors can be played off against each other through coalition building to achieve political results.

Leggett embarked in September 1992 on a tour of European, American, and Japanese insurers and reinsurers, attempting to persuade them of the need to move beyond a sense of alarm at the rise in catastrophe payouts and towards a proactive engagement with climate politics. In the end, as he concedes in his book, he was only able to help bring about a small, committed coalition of activists within particular companies, almost exclusively European.[8] These activists attended the workshop Leggett organized at the first Conference of the Parties to the UNFCCC in Berlin in 1995, contributing presentations later collected and published (Leggett 1996, 326). United Nations Environment Program (UNEP) got involved in 1995, with the development of a Statement of Environmental Commitment by the Insurance Industry (UNEP 1995). The core members from the insurance industry involved in Leggett's 1995 workshop were all also involved in this, and remain the coordinating group of UNEP's initiative. In 1997, UNEP established its Insurance Industry Initiative, which "funds research activities, sponsors awareness meetings and workshops and the annual regular meetings of the Initiative" (Bode 2000). UNEP has tried clearly to use the insurers to help build a coalition in favor of action to mitigate climate change, and also to develop a variety of means, from workshops on investment best practice, to the development of a greenhouse gas benchmark designed to enable investors to translate the greenhouse gas (mainly CO_2) intensity of a company into a financial risk and thus deter investment in companies that are energy inefficient (Thomas et al. 2000).

One of the last results of the UNEP and insurance industry collaboration is the report *The Kyoto Protocol and Beyond: Potential Implications for the Insurance Industry* (Salt et al. 1998),[9] which presents the prospects and challenges to the industry with respect to climate-related issues. In the report, space is opened up for further collaboration with state-actors by arguing that the insurance industry has acquired knowledge and experience useful to the rest of society. Examples of areas for fruitful collaboration are the planning and implementation of disaster mitigation, which could lead to more effective systems of risk management, as well as enhanced modelling of extreme events. Suggestions to physically alter and lessen climate risk are seldom discussed during the negotiation rounds. The political negotiations have instead dealt with emission reductions as a means to eliminate greenhouse gases. However, risks may be decreased not only by preventing them from emerging, but also by managing their consequences. Discussions on how to handle the effects of climate change have, however, been rare so far. One reason is its politically complicated and sensitive nature, especially from a global perspective; it raises not only the issue of responsibility but also the question of who pays for such measures (Carlsson and Stripple 1999).

Furthermore, although the climate change issue is surrounded by scientific uncertainties, the insurance industry has enlarged its collaboration with atmospheric scientists to increase the predictive accuracy of catastrophe modelling (Keykhah 2000; Paterson 2001). While scientific uncertainty is usually used by different actors to legitimate a "wait-and-see" approach toward policymaking, the industry has shown an outspoken support for the precautionary principle as a guidance for decisions in the face of scientific uncertainty. This support may have a symbolic importance as it draws on the industry's experience in confronting and managing uncertainty and risk. It also implies that a principle considered important in environmental policymaking has gained general support from a major industry.

While there was a flurry of high-profile activity between 1995 and 1998 by this industry group, the activists proved unable, by and large, either to expand the range of companies closely involved in either the work of UNEP's initiative, or the status of the commitment by their companies to mitigating climate change. Some of them, for example Swiss Re, have developed portfolios of investment into renewable energy, having invested CHF46 million (approx. $30 million) by the end of 2000 in such projects (Swiss Re 2000). New companies have been formed to channel funds from institutional investors into renewable projects, notably Leggett's own Solar Century, but also Sustainable Asset Management, and New Energy Invest. But the level of such investment by insurers is still tiny as a proportion of their overall (or even new) investments, and the number of companies actively involved is still made up mostly of those who were involved in the early 1990s. Most companies remain unconvinced that it is possible for them to shape energy futures and thus climate impacts through their investment practices, even if many of them are concerned about the increase in, and especially the non-linear increase in, catastrophic weather events.

The Securitization of Insurance Risks

As suggested previously, while many insurers, and particularly reinsurers, have been worried about the implication of global warming for their industry, only a few have felt that the appropriate response was to think about their investment practice by promoting the development of renewable energy to mitigate global warming, and by extension long-term patterns in the incidence of severe weather. In this sense, the Greenpeace/UNEP alliance building was successful to the extent that the insurance industry grew concerned, but they did not really become engaged in the type of activities that the Greenpeace/UNEP envisaged. Insurers, instead, have principally adopted financial strategies to limit their exposure to risk. On the one hand, the strategy of collaborating with environmentalists had significant costs for insurers, both politically and economically. We have dealt elsewhere with the reasons why insurers have not been able to develop investment strategies that Leggett and others would have wished. Explanations include the interrelations between insurers as investors and the manufacturing firms they invest in, which would be affected by CO_2 reduction strategies; ownership of insurers by such manufacturing firms (e.g., through their pension funds); limits on investment switching strategies because insurers are often price-makers in stock markets due to their financial size; a culture of nonintervention in corporate decisions, or passive investment; and regulatory arrangements limiting types of investment insurers may make (Carlsson and Stripple 1998; Paterson 2001). According to representatives of the Swedish insurance industry, one important reason was that these measures were "seen as too far-fetched to attract investors" and more full-blooded attempts to get involved in the climate politics would risk jeopardizing "the responsibility to the stockholders" (Carlsson and Stripple 1998, 341). But perhaps more importantly, new opportunities emerged for insurers that enabled them to limit their exposure to risk using purely financial mechanisms familiar to them.

There have been some other responses, with certain firms and insurers' organizations engaging in lobbying governments to act to limit greenhouse gas emissions, most looking closely at premiums and deductibles for catastrophe cover to limit their exposure to risk, and some collaborating with atmospheric scientists as already noted. But the response we want to focus on is the emergence of new financial markets in Insurance Linked Securities (ILS), as a new set of financial mechanisms through which direct risk is reinsured. This is important partly because we think it is becoming a dominant response by the industry, but also because it enables us to think about practices of governance by business in global environmental politics.

The securitization of insurance risks is a new way of spreading the costs of catastrophes. Securitization is a process that turns an insurance risk into a marketable security. So the basic idea is to transfer insurance risks, not to the insurance market, but to the capital market where the risks are assumed by private or institutional investors instead of by insurers and reinsurers. Securitization implies that the risk, for example, of a hurricane in a specific place at a defined magnitude or loss, is packaged in a standardised form, for example, as a bond, or an option or futures contract, and sold on the capital market. Hence, the risk is "secured" on the capital market. At the same time, securitization of insurance risks holds a number of benefits for potential investors. The investor who buys the instrument issued by the insurer will lose his/her money if there is a catastrophe of a defined magnitude of loss. If there is no catastrophe, the investor will get their money back with an attractive rate of interest. But perhaps more importantly, the risk is not correlated with the traditional market risks associated with normal capital market investments. This makes risk-transfer mechanisms such as catastrophe bonds attractive for investors who want to hold a diversified portfolio.

In December 1992, catastrophe futures first began to be traded on the Chicago Board of Trade (Punter 1999; Stix 1996, 28; Tynes 2000, 7). Formally known as a set of financial instruments called Insurance-Linked Securities, the rationale for reinsurers was that as payouts on large-scale weather disasters increased, the available pool of money to pay out would be insufficient using traditional reinsurance techniques (Chookaszian 1998; Michaels et al. 1997). For example, the total pool of catastrophe capital in the United States is $6–8 billion, and the total pool of the reinsurance industry only $30 billion. By contrast, events with the intensity of Hurricane Andrew could cause losses of up to $100 billion (Michaels et al. 1997). Globally, the size of the insurance market is estimated at around $2000 billion, the reinsurance market holds another $200 billion, while the capital market holds $20,000 billion (Salt et al. 1998).

However, reinsurers could access the much larger pool available through the capital markets, thus making the system considerably less vulnerable to a quick succession of disasters with losses such as occurred in Hurricane Andrew. At the same time, the high payouts in the late 1980s and early 1990s drove the price of traditional reinsurance up, meaning there might be demand for different forms of reinsurance that might be cheaper (Swiss Re 2001, 3; Tynes 2000). The first instruments traded were futures and options contracts, but from 1995–1996 onward, catastrophe bonds were developed, and quickly became more successful. These are instruments where the purchaser of the bond agrees to forfeit the profits that would be made (and often even their principal) should a specified event occur. In return, high rates of return are available and, importantly for investors, they diversify risk since the risk of catastrophes is not correlated with those of the normal risks of investments in financial markets (Cabral 1999; Tynes 2000). Effectively, this is a renewal of an older form of insurance (so-called "debt forgiveness") where as long ago as Hammurabi, King of Babylon, ships were insured via advances from lenders who agreed to forfeit their loan should the ship be lost at sea (Punter 1999).

Since 1997, these sorts of contracts have grown rapidly (Punter 1999). According to Swiss Re, roughly US$8 billion of such financial instruments had been issued between 1996 and early 2001 (Swiss Re 2001, 3). The pace slowed in 1999 and 2000 as the price of traditional reinsurance declined (due to a fall in catastrophe payouts), but picked up again in 2001 due to a number of catastrophes in 2000 (Swiss Re 2001, 18, 32).The possibility of engaging in securitization of insurance risks has been enabled by new developments in meteorology from about 1994 onward. These have made possible much more accurate assessments of the likelihood of severe weather events in particular areas over a medium time frame, up to eighteen months (the period of property insurance is, of course, normally one year) (e.g., Land 1996, Nutter 1996).

Although the development of such instruments is still in its infancy, it is clear from the prominence of discussions of this in online journals such as *Global Reinsurance*, that they are regarded by reinsurers and the financial markets as a promising development, both in terms of stabilizing and managing risk for insurers and reinsurers, and as lucrative markets for investors. Tynes reports that several reinsurers (including Swiss Re and Hanover Re) as well as investment banks such as Goldman Sachs and Merrill Lynch are competing to take advantage of financial markets by developing and issuing catastrophe-related financial instruments (Tynes 2000, 7). However, as noted before, one of the key conditions that have made such developments possible has been the increasing credibility of long-range models for predicting the likelihood of natural disasters, giving investors a sense of the risk they are undertaking (Punter 1999); hence the interest that some climatologists have taken in marketing their services to reinsurers, as suggested earlier. Such a development would, in principle, mean that even if long-term climate changes were occurring, the insurance business has the capacity to make its own operations secure and stable.

Privatized Governance

A number of authors emphasize that (inter)state governance increasingly serves corporate needs. Such arguments have been made widely in general terms (e.g., Hertz 2001) as well as specifically in relation to the UNCED process (Chatterjee and Finger 1994; Sachs 1993), as well as other multilateral economic and environmental processes such as the WTO (Conca 2000), the TRIPS and TRIMS provisions in the Uruguay round of the GATT (Purdue 1996; Sell 1999), the MAI (Egan and Levy 2001; Picciotto 1999), and NAFTA (MacArthur 2001; Nader et al. 1993; Rupert 2000, 65–93). They have also been made in the context of individual states, for example, in Monbiot's account of the "captive state" in the UK (Monbiot 2000).

But at the same time, and perhaps more importantly in terms of shifts in global political structures and processes, corporations are increasingly not merely playing a role in the negotiation and maintenance of public and state-based regimes, but are also in many cases establishing "private regimes." Interfirm global governance bypassing states is on the increase. Clapp (1998), for example, shows how the International Organisation for Standardisation (ISO) produces regulations on business practice that have become a form of global business self-regulation. Rather than states seeking to regulate business practice in order to pursue social, political, or environmental goals, businesses are now enabled to set such standards themselves.

In other fields, Timothy Sinclair (1999) shows how bond-rating agencies, such as Moody's or Standard and Poor's, are increasingly able to practice governance on firms and states by shaping the conditions under which investors will invest in companies. The financial community accepts the role of the rating agencies and awards them the authority to judge their behavior. The market actors anticipate the views and expectations of the rating agencies on what is regarded as acceptable. According to Sinclair (1999, 161) these agencies, as well as other types of coordination services firms, do not merely constrain capital markets but contribute to their constitution as agents. In this sense, the rating agencies have a form of structural power that not only produces the outcomes but also the subjects.

While Sinclair and Clapp have provided contemporary examples of privatization of governance, Claire Cutler (1999) offers a longer historical perspective in her account of private international trade law and, in particular, the laws and practices that govern maritime transport. Cutler shows that merchants and their associations have emerged as "private authorities" with a capacity to exercise influence on maritime transport laws. According to liberal thought, private relations are seen to be "apolitical" in content and effect and the law governing those relations is seen to be operating neutrally among participants in the market. However, Cutler asserts that private international trade law has a much more political nature and history. It works in a distributional way to determine who gets what, when, and how. In the field of maritime transport, private authority operates protectively to preserve the influence of the most powerful maritime states. On a more general level, she argues that the distinction between public and private law creates obstacles to theorizing about the political nature of private actors and to conceptualizing private action as political action. Finally, Virginia Haufler (1997a, 1997b) has indicated how, in other aspects of the field we discuss here, insurance companies have been able to define the sorts of business and political risks other firms can undertake. She has principally shown this in the field of political and particularly war risks insurance, demonstrating how the norms, which mostly have emerged from Lloyd's of London, have been diffused throughout the industry and served to discipline business and governments alike.

In some instances, authors have also shown how privatized forms of governance have involved alliances between environmental nongovernmental organizations and private firms, in which environmental NGOs can achieve some influence. Peter Newell's work is perhaps the most fully developed here (Newell 2001). Newell gives a wide range of examples of types of relationships that have emerged between environmental NGOs and firms—from shareholder activism where NGO members buy shares in companies to put pressure on them, through to boycott campaigns, to more constructive engagement such as the fostering of labelling schemes (Newell 2001, 94). Such practices create systems of incentives and "soft sanctions" that serve to shape the conduct of firms. Wapner's work on the Coalition for Environmentally Responsible Economies (CERES) principles (1997) and other work on the Forestry Stewardship Council (FSC) displays a similar logic—patterns of privatized governance emerging where nongovernmental but noncorporate actors regulate the practices of private firms (e.g., Humphreys 1996; Murphy and Bendell 1997).

Securitization as Global (Privatized) Governance: Triage

To reiterate, the strategies of insurers should not be understood as simply economic actions by rational firms with no political implications. Rather, insurance and securitization ought to be fundamentally seen as practices of governance. Normatively, they regulate the behavior of others in a variety of ways. This therefore represents a privatized form of governance whereby insurance companies arrogate to themselves the decisions about who is to be made subject to large-scale climate related risks.

The emergence of new financial instruments—catastrophe bonds, options, futures, and the like—creates opportunities for new forms of market behavior by insurers (and investors) and the insured to secure their own risks in the context of increasingly uncertain weather patterns and heightened physical insecurity. But at the same time, securitization has developed alongside the redistribution of the risks of weather related disasters. Securitization is, of course, itself a redistribution of such risks among the investors prepared to undertake such financial risks. But as a process, it has accompanied a noticeable reduction in cover in some areas prone to hurricanes or cyclones, either directly through withdrawal of cover, or through rate increases that mean that some "choose" (or more likely are simply unable) to pay for insurance.

Leggett notes a number of such instances. Early on in his attempts to persuade insurers of the need to become involved in global warming politics, he visited a syndicate at Lloyd's. The immediate result of Hurricane Andrew was the increased cost of catastrophe reinsurance, and reduced coverage. The syndicate he visited had in fact commissioned a report from a UK climate research center about global warming, which predicted increased hurricanes in Florida. The syndicate withdrew cover from the region prior to Hurricane Andrew, thus saving itself millions of dollars (Leggett 1999, 104). Leggett was also told that "insurance cover had already been withdrawn in many parts of the Caribbean", and Leggett notes that he already knew coverage had also been withdrawn in Western Samoa (p. 105). Actually, the high global incidence of extreme weather events in the early 1990s generated significant reinsurance shortages. Even in those areas where insurance was still available, in Caribbean countries the insurance rate increased by 200 percent–300 percent (Pollner 2001).

After large losses there, the Hawaiian Insurance Group had to cease trading, and the largest insurer in Hawaii, First Insurance, announced it would not renew existing policies as of February 1993. The latter action left 38,000 homes uninsured (Leggett 1999, 117). Further such announcements were made in April 1993, leading Leggett to refer to what he called a "climatic domino effect . . . First the insurers felt obliged to withdraw from Pacific and Caribbean island nations. Then, after Hurricane Andrew and Cyclone Iniki, they had felt obliged to try to withdraw from Hawaii and Florida. Then would come coastal areas more generally. And so on" (1999, 121). But excepting some effects in Hawaii, by and large the domino stopped at the doors of the "industrialized world." In some cases, notably Florida and Japan, states managed to prevent withdrawal of cover through various regulatory measures. But in addition, the income gained from insuring industrialized countries is such that full-scale withdrawal of cover from such areas is unlikely to be financially viable.

Thus, the withdrawal of cover mirrors patterns of power in the global political economy fairly closely. Even when cover is withdrawn from those in the industrialized world, this is more likely to occur because of rate increases and thus self-exclusion (according to patterns of economic inequality within industrialized countries)—a well-known process of "financial exclusion," which exists in other aspects of insurance as well as in banking, especially in the Anglo-American neoliberal heartland (Leyshon and Thrift 1997, 222–259).

But there is a certain amount of novelty here in the way in which insurance acts in exclusionary ways across borders. In an analysis of insurance in terms of Foucault's notion of governmentality, Ewald (1991) shows, as in the epigram at the beginning of the chapter, that insurance can serve to produce and reinforce social inequalities. But he also shows that insurance can produce community and solidarity. Insurance "constitutes a mode of association which allows its participants to agree on the rule of justice they will subscribe to" (Ewald 1991, 207). He argues that insurance is primarily to be thought of as a "certain type of rationality" rather than a specific institution (Ewald 1991, 206). Such a rationality operates to render complex processes, such as global warming, into calculable and thus manageable risks. Such a construction aggregates risks into a collective whole, but simultaneously differentiates subjects into those deemed "safe" and "risky." Such a rationality tends, of course, to obscure the origins of risks, instead focusing on governing the individuals and collectivities on which risks are placed.

But what appears through Ewald's analysis of insurance as a form of rationality, especially reflecting on it in the global warming context, is that the contexts where insurance appears as "a new privilege for the rich and a cruel irony for the poor" is in terms of commercial insurance, where the community is primarily formed out of commercial ties and thus the members are generally affluent. By contrast, more "universal," solidaristic experience of the possibilities of insurance appear in the forms of social insurance such as unemployment and pensions insurance. In the global warming context, both forms have exclusionary consequences. For although the "universalistic" form of insurance might help to constitute forms of solidarity, its universalism is strictly limited to national boundaries—they are the product of the emergence of the welfare state in Western societies since the early twentieth century.

No such corresponding solidarities exist internationally. In the international system, the victims and their governments bear the major share of the losses from natural disasters, that is, there is little global sharing of natural disaster losses (Linneroth-Bayer and Amendola 2000). Further, even if the forms of catastrophe insurance that emerge in the context of global warming were to be organized through states, though this now seems unlikely, these would not, in the context of a global problem like climate change, provide in any meaningful sense a "universal" form of security. But conversely, insurance in its commercial mode simultaneously produces exclusions of its own, through the crude mechanisms of "ability to pay." In this sense, the climate change issue and the politics of securitization raises fundamental issues of what it means to be secure, and how security should be organized.

The securitization of risks as a form of privatized governance thus has significant justice implications. Especially absent measures to mitigate CO_2 emissions (which insurers could contribute significantly to, see Paterson 2001), securitization of risks enables those in the rich West to secure themselves against the global warming that they are primarily responsible for, while those in the global South are exposed to increased risks, and are largely, and increasingly, uninsured (as well as more frequently living and working in buildings which are more vulnerable in the face of extreme weather events). Thus while those suffering Hurricane Andrew will be (relatively) secure, those experiencing

Hurricane Mitch, or the Orissa floods, are made more vulnerable. The background condition to this is, of course, that natural disasters are not evenly distributed over the world, and that they are in fact not that "natural" after all.

Poverty is generally recognized as one of the most important correlates of vulnerability to natural disasters. Other correlates include differences in health, gender, ethnicity, and education (Dow 1992; Liverman 1990; quoted in Meyer et al. 1998, 240). On a global scale this is exemplified by far higher losses of life due to similar extreme events in the developing world than in the developed world (Meyer et al. 1998). In terms of economic costs, in absolute numbers the direct costs of natural catastrophes are evenly split between the developed and the developing world. However, given the dramatic differences in GDP, the per capita cost of natural disasters in relation to GDP is significantly higher for developing countries (Freeman 1999). In that sense, the extent to which the weather is "extreme" is not only dependent on where one is situated geographically, but also one's location in economic and political structures.

The pattern of governance practiced by insurers thus sits well with the accounts of privatisation in global governance discussed previously. The process of governance of global warming by insurers mirrors most closely Clapp's (1998) argument about privatized governance, albeit through rather different mechanisms. Clapp suggests that the privatization of standard setting in the ISO is contributing to increased inequalities as it excludes many in developing countries from being able to compete for investment. Here, privatized governance creates injustices by creating heightened risks for all, while enabling some to insure themselves against (the financial aspects of) those risks. It selects between those deemed worth securing against the heightened risks of climate change, and those who should be left to their own devices: it practices triage.

Technically, triage is the technique developed in military hospitals to distinguish between those who will die anyway and who thus should be ignored, those who will survive but do not require urgent care, and those for whom immediate medical intervention is most likely to increase the chances of survival. A category of nurses in many countries in A&E/ER

departments is still called "triage nurses." But in global environmental politics, the phrase was used widely in the 1970s as a charge against Garrett Hardin's "lifeboat ethics" (1974), where he claimed that the "industrialized countries," in the context of environmental and (putative) population crises, should pull up the ladder and ignore the fate of the "developing" world. Insurers, in a limited but important way, may be in the process of doing Hardin's bidding.

Conclusions

In this chapter we have tried to show the interconnection between two political processes. On the one hand, elements within the environmental movement (Greenpeace, UNEP) attempted to construct a coalition with insurers in a process that could be termed, in a limited sense at least, a counterhegemonic project. This project has not attempted to construct a broad coalition against the dominant neoliberal historic bloc, but rather simply to split that bloc in relation to climate change politics. But at the same time, the neoliberalism on which the dominant bloc has secured its hegemony has created possibilities for the adoption of various practices that may be termed privatized governance. These enable insurers to avoid having to negotiate or ally with environmentalists or states advocating GHG cuts, or to develop measures to mitigate CO_2 emissions. Thus what has occurred in this field is a form of privatized governance, which is acting to redistribute significantly the risks associated with the impacts of climate change in highly inegalitarian ways. As a result, Greenpeace has ultimately been unable to project a conception of the general interest (understood as GHG emissions reductions) in a manner sufficient to incorporate insurers into the challenge to the dominant fossil fuel industry. Insurers have been willing to advocate that governments act to cut emissions, but they have not, by and large, been particularly active in the political realm in pursuit of this objective, nor have they even reorganised their own investment practices to this end.

While the construction of hegemonic projects generally involve bargaining, Levy and Newell suggest in chapter 3 of this book that bargains tend to be unequal, and usually privilege capital systematically. In the case presented here, business has successfully preserved its privileged position almost precisely by constructing arenas for political action (capital markets) where no bargaining with other actors is required. Greenpeace and UNEP may have been able to foster some discursive changes in insurance practice, for example with their adoption of the precautionary principle and broad support for abatement action, and some organizational changes in terms of their participation in the UNEP Insurance Industry Initiative. But the operations of the economic system have undermined the successful construction of a counterhegemonic project in climate politics. The emergence of market-based solutions, in particular the securitization of climate-related insurance risks, has been an exit option for insurers that has minimized their need to involve themselves in explicitly political processes surrounding climate change.

Notes

1. This chapter draws on earlier work by all three authors. See Paterson 1999, 2001; Carlsson and Stripple 1998, 1999; Jagers and Stripple 2003; Stripple 1998, 2000; Stripple, Chong, and Wiman 1997. (Note that Sverker Jagers was formerly known as Sverker Carlsson).

2. See for example also ABI (2000), for evidence that insurers in general are very clear and vocal in their support for neoliberalism.

3. The principal greenhouse gas (GHG) is carbon dioxide, but gases such as methane, nitrous oxide, and chlorofluorocarbons (CFCs) are also GHGs.

4. The Intergovernmental Panel on Climate Change is a specialised intergovernmental body established by the United Nations Environmental Programme (UNEP) and World Meteorological Organization (WMO) in 1988 to provide assessments of the results of climate change research for policymakers. The IPCC, which consists of over 2,000 scientists from around the world, has compiled three "Assessment Reports" (1990, 1995, 2001).

5. Figures in constant 1998 U.S. dollars.

6. Figures in constant 1990 U.S. dollars. According to Susan Strange, it also contributed to concentration in the international reinsurance industry, as firms merged to consolidate risks or left the business to avoid exposure to such catastrophic risks (Strange 1996, 132–133).

7. This account draws largely on Leggett's own depiction of his involvement in climate politics in his book, *The Carbon War* (1999). It is thus inevitably a little skewed. But the account is not, in our view, inconsistent with journalistic commentaries throughout the period, or the direct observation of climate politics by the authors.

8. The most well known of these have been Andrew Dlugolecki of General Accident (now CGNU), Tessa Tennant then of NPI, Carlos Joly then at Storebrand in Norway, Ivo Knoepfel of Swiss Re, Rolf Gerling of Gerling Konzern, and Frank Nutter of the Reinsurance Association of America.

9. Julian Salt works for the Loss Prevention Council in London, an institute of the Association of British Insurers.

References

ABI (2000). *Annual report 1999–2000*. London: Association of British Insurers.

Bode, A. (2000). Frequently asked questions about the UNEP insurance industry initiative. Geneva: United Nations Environment Programme. Available at <http://www.unep.ch/etu/finserv/insurance/question.htm>.

Cabral, W. (1999). Securitisation out of the niche. *Global Reinsurance, 8*, 1. Online journal available at <http://www.globalreinsurance.com>.

Carlsson, S., and Stripple, J. (1998). Climate change and the insurance industry's response. *Scandinavian Insurance Quarterly, 4*, 335–344.

Carlsson, S., and Stripple, J. (1999). The insurance industry and the climate issue: Pre and post Kyoto, in perspective. *Scandinavian Insurance Quarterly 4*, 351–362.

Carlsson, S., and Stripple, J. (2000). Climate governance beyond the state; Contributions from the insurance industry. Paper presented at the XVIIIth World Congress of Political Science (IPSA), Québec City, August 1–5, 2000.

Chatterjee, P., and Finger, M. (1994). *The earth brokers: Power, politics and world development*. London: Routledge.

Chookaszian, D. (1998). Securitisation: The first five years. *Global Reinsurance, 7*(5), Online journal available at <http://www.globalreinsurance.com>.

Clapp, J. (1998). The privatization of global environmental governance: ISO 14000 and the developing world. *Global Governance, 4*, 295–316.

Conca, K. (2000). The WTO and the undermining of global environmental governance. *Review of International Political Economy, 7*(3), 484–494.

Cutler, C. A. (1999). Private authority in international trade relations: The case of maritime transport. In C. A. Cutler, V. Haufler, & T. Porter (Eds.), *Private authority and international affairs* (pp. 283–329). Albany, NY: State University of New York Press.

Dlugolecki, A. (1996). An insurer's perspective. In J. Leggett (Ed.) *Climate Change and the Financial Sector* (pp. 64–81). Munich: Gerling Akademie Verlag.

Dow, K. (1992). Exploring differences in our common future(s): The meaning of vulnerability to global environmental change. *Geoforum, 23*, 417–436.

Egan D., & Levy, D. (2001). International environmental politics and the internationalization of the state: The case of climate change and the multilateral agreement on investment. In D. Stevis & V. Assetto (Eds.), *The international political economy of the environment: Critical perspectives IPE Yearbook 12* (pp. 63–83). Boulder, CO: Lynne Rienner.

Ewald, F. (1991). Insurance and risk. In G. Burchell, C. Gordon, & P. Miller (Eds.), *The Foucault effect: Studies in governmentality* (pp. 211–233). Chicago: University of Chicago Press.

Freeman, P. (1999). Infrastructure, natural disasters, and poverty. Laxenburg, Austria: International Institute for Applied Systems Analysis (IIASA), draft paper.

Hardin, G. (1974). Lifeboat ethics: The case against helping the poor. *Psychology Today, 8,* pp. 38–43.

Harmes, A. (1998). Institutional investors and the reproduction of neoliberalism. *Review of International Political Economy, 5,* 92–121.

Haufler, V. (1997a). *Dangerous commerce: Insurance and the management of international risk.* Ithaca, NY: Cornell University Press.

Haufler, V. (1997b). Financial deregulation and the transformation of International Risks Insurance. In G. Underhill (Ed.), *The new world order in international finance* (pp. 76–100). London: Macmillan.

Hertz, N. (2001). *The silent takeover: Global capitalism and the death of democracy.* London: Random House.

Humphreys, D. (1996). *Forest politics: The evolution of international cooperation.* London: Earthscan.

IPCC. (1990). *IPCC first assessment report.* Three volumes. Cambridge: Cambridge University Press.

IPCC WG I (1995). *Climate change 1995: The science of climate change—Contribution of Working Group 1 to the Second Assessment Report of the Intergovernmental Panel of Climate Change.* Cambridge, UK: Cambridge University Press.

IPCC WG I (2001). *Climate change 2001: The scientific basis—Summary for policymakers: A Report of Working Group 1 of the International Panel of Climate Change.* Shanghai: World Meteorological Organization/United Nations Environment Programme.

IPCC WG II (1995). *Climate Change 1995: Impacts, Adaptations and Mitigations of Climate Change—Scientific-Technical Analyses, Contribution of Working Group II to the Second Assessment Report of the Intergovernmental Panel of Climate Change.* Cambridge, UK: Cambridge University Press.

Jagers, S. C., and Stripple, J. (2003). Climate governance beyond the state. *Global Governance, 9,* 385–399.

Kerr, R. A. (2000). Draft report affirms human influence. *Science, 288,* 58–59.

Keykhah, M. (2000). *Global hazards and catastrophic risk: assessments, practitioners, and decision making in reinsurance.* Belfer Center for Science and International Affairs (BCSIA) Discussion Paper 2000-22, Environment and Natural Resources Program, Kennedy School of Government, Harvard University.

Krause, K., & Williams, M. C. (1997). From strategy to security: Foundations of critical security studies. In K. Krause & M. Williams (Eds.), *Critical Security Studies* (pp. 33–60). Minneapolis: University of Minnesota Press.

Kron, W. (2000). Natural disasters: Lessons from the past—Concerns for the future. *Geneva Papers on Risk and Insurance: Issues and Practice, 25*(4), 570–581.

Land, T. (1996). Long-term weather forecasting. *Contemporary Review, 269*(1566), 21–24.

Leggett, J. (1996). *Climate change and the financial sector; The emerging threat—The solar solution.* München: Gerling Verlag.

Leggett, J. (1999). *The carbon war: Dispatches from the end of the oil century.* London: Penguin.

Leggett, J. (no date). *Climate change and the insurance industry: Solidarity among the risk community?* London: Greenpeace.

Leyshon, A., & Thrift, N. (1997). *Money/Space.* London: Routledge.

Linneroth-Bayer, J., & Amendola, A. (2000). Global change, natural disasters and loss sharing: Issues of efficiency and equity. *The Geneva Papers on Risk and Insurance, 25*(2), 203–219.

Liverman, D. M. (1990). Vulnerability to global environmental change. In R. E. Kasperson, K. Dow, D. Golding, & J. X. Kasperson (Eds.), *Understanding global environmental change: The contributions of risk analysis and management* (pp. 17–44). Worcester, MA: ET Program, Clark University.

MacArthur, J. R. (2001). *The Selling of "Free Trade": NAFTA, Washington, and the Subversion of American Democracy.* Berkeley: University of California Press.

Meyer, W. B., Butzer, K. W., Downing, T. E., Turner, B. L., Wenzel G. W., & Wescoat, J. L. (1998). Reasoning by analogy. In S. Rayner & E. Malone (Eds.), *Human choice and climate change* (pp. 217–289). Columbus, OH: Battelle Press.

Michaels, A., Malmquist, D., Knap, A., & Close, A. (1997). Climate science and insurance risk. *Nature, 389*(6648), 225–228.

Monbiot, G. (2000). *Captive state: The corporate takeover of Britain.* London: Macmillan.

Munich Reinsurance (1998). *Topics 1998.* Munich: Munich Reinsurance.

Murphy, D. F., & Bendell, J. (1997). *In the company of partners.* Bristol: The Policy Press.

Nader, R., Greider, W., & Atwood, M. (Eds.) (1993). *The case against free trade: Gatt, NAFTA and the globalization of corporate power.* Washington, DC: Earth Island Press.

Newell, P. (2001). Environmental NGOs, TNCs, and the question of governance. In D. Stevis & V. Assetto (Eds.), *The International political economy of the envi-*

ronment: Critical perspectives. IPE Yearbook 12 (pp. 85–107). Boulder: Lynne Rienner Publishers.

Nutter, F. W. (1996). A reinsurer's perspective. In J. Leggett (Ed.), *Climate change and the financial sector* (pp. 82–90). Munich: Gerling Akademie Verlag.

Oberthur, S., & Ott, H. E. (1999). *The Kyoto protocol: International climate policy for the 21st century.* Berlin: Springer Verlag.

Paterson, M. (1999). Insurance companies and the politics of global warming. *IDS Bulletin, 30*(3), 25–30.

Paterson, M. (2001). Risky business: Insurance companies in global warming politics. *Global Environmental Politics, 1*(4), 18–42.

Picciotto, S. (1999). A critical assessment of the MAI. In S. Picciotto & R. Mayne (Eds.), *Regulating International Business* (pp. 82–108). London: Macmillan.

Pollner, J. D. (2001). *Managing catastrophic disaster risks using alternative risk financing and pooled insurance structures.* Washington, DC: The World Bank, Finance, Private Sector and Infrastructure Department. Finance, Private Sector and Infrastructure Department.

Punter, A. (1999). Insurance-linked securitisation: A progress report. *Global Reinsurance* 8 (8). Online journal available at <http://www.globalreinsurance.com>.

Purdue, D. (1996). Hegemonic trips: World trade, intellectual property and bio-diversity. *Environmental Politics, 4*(1), 88–107.

Rupert, M. (2000). *Ideologies of globalization: Contending visions of a new world order.* London: Routledge.

Sachs, W. (Ed.) (1993). *Global ecology.* London: Zed.

Salt, J., Knoepfel, I., Bode A., & Jakobi, W. (1998). *The Kyoto protocol and beyond: Potential implications for the insurance industry.* UNEP: UNEP Insurance Industry Initiative.

Schmidheiny, S. (Ed.) (1992). *Changing course.* Cambridge, MA: MIT Press.

Schmidheiny, S., & Zorraquin, F. J. L. (1996). *Financing change: The financial community, eco-efficiency and sustainable development.* Cambridge, MA: MIT Press.

Sell, S. (1999). Multinational corporations as agents of change: The globalisation of intellectual property rights. In C. A. Cutler, V. Hanfler, & T. Porter (Eds.), *Private Authority and International Affairs* (pp. 169–198). Albany: State University of New York Press.

Sinclair, T. (1999). Bond-rating agencies and the coordination in the global political economy. In C. A. Culter, V. Hanfler, T. Porter, *Private authority and international affairs* (pp. 153–168). Albany, NY: State University of New York Press.

Skjaerseth, J. B., & Skodvin, T. (2001). Climate change and the oil industry: Common problems, different strategies. *Global Environmental Politics, 1*(4), 43–64.

Stix, G. (1996). Green policies. *Scientific American, 274*(2), 27–28.

Strange, S. (1996). Insurance business: The risk managers. In *The retreat of the state* (pp. 122–134). Cambridge University: Cambridge University Press.

Stripple, J. (1998). *Securitizing the risks of climate change: Institutional innovations in the insurance of catastrophic risk*, IIASA Interim Report IR-98-098/December. Laxenburg, Austria: International Institute for Applied Systems Analysis.

Stripple, J. (2000). The sensitivity of the international insurance industry to climate change. In B. Wiman, J. Stripple, & S. M. Chong (Eds.), *From climate risk to climate security* (pp. 53–64). Lund, Sweden: Kalmar University, Sweden and Swedish Environmental Protection Agency.

Stripple, J. (2001). *Climate change and international relations; Reconsidering interdependence, governance and security.* Licentiate Thesis, Department of Biology and Environmental Science, University of Kalmar, Kalmar.

Stripple, J., Chong, S., & Wiman, B. L. M. (1997). Turning global environmental risks into sustainable societal resource management: A study of institutional responses. Project proposal Summary and Description, Lund Sweden: Lund University, Sweden.

Swiss Re. (2000). *Environmental report 2000.* Zurich: Swiss Re. Available at <http://www.swissre.com>.

Swiss Re. (2001). *Capital market innovation in the insurance industry.* Sigma Report No 3/2001. Zurich: Swiss Re.

Thomas, C., Tennant, T., & Rolls, J. (2000). *The GHG indicator: UNEP guidelines for calculating greenhouse gas emissions for business and non-commercial organisations.* Geneva: United Nations Environment Programme.

Tynes, J. S. (2000). Catastrophe risk securitization. *Journal of Insurance Regulation, 19*(1), 3–28.

UNEP (1995). *Statement of environmental commitment by the insurance industry.* Geneva: United Nations Environment Programme. Available at <http://www.unep.ch/etu/finserv/insurance/statemen.htm>.

Wapner, P. (1997). Governance in Global Civil Society. In O. Young (Ed.), *Global governance: Drawing insights from the environmental experience* (pp. 65–84). London: MIT Press.

11

The New Water Paradigm: The Privatization of Governance and the Instrumentalization of the State

Matthias Finger

Introduction

This chapter addresses the changing relationships among business, the state, civil society, and international institutions, taking water and its management as a paradigmatic case. More precisely, it makes the argument that—at least in the case of water—the state is increasingly becoming instrumentalized as part of a new global institutional framework, to create new business opportunities for transnational corporations. This, I will argue, is a two-stage process, in which environmental concerns, other threats to human security, as well as civil society involvement have actively contributed to the outcome. The argument is illustrated with case studies from the water supply and sanitation sectors, or the so-called "water sector."

Conceptually, my approach is essentially organizational and institutional in nature: actors, who all pursue their own more or less well articulated interests, operate within layers of institutional frameworks (Giddens 1984). The concept of "institutions" is used here in a broad sense and includes all relevant societal actors, in particular government, business, and civil society, as well as the rules and norms governing their relationships. Institutions are viewed as codified power structures that shape the behavior of the actors involved. Simultaneously, institutions are shaped by these actors as they compete to exert agency over institutional frameworks, ideologies, and practices.

By forming strategic alliances, the different actors may be capable of reshaping the institutions within which they operate (Crozier 1963). The context of globalization of the 1990s offered a unique opportunity for the various actors to redefine the institutional "rules of the game." In the water sector, the transnational corporations have managed to take advantage of the recent globalization process in order to redefine the power relationships among actors to their advantage, codifying furthermore this new power relationship in a new institutional framework and corresponding ideology.

This chapter is divided into two main parts. The first examines the different factors that have led to the emergence of what I call a "new water paradigm," including the rise of environmental issues on the international agenda, and the increasing emphasis on market mechanisms in water management at all levels, from local to global. The legitimation of such market mechanisms has, in turn, made private sector participation in the water supply and sanitation sector more acceptable. In particular, I look at the historical process through which civil society organizations have emerged, partnerships have been formed at multiple levels, and alliances have been constructed involving all kind of actors, from local NGOs to transnational corporations, nation-states, international financial institutions, United Nations agencies, and international NGOs. In the second part, I analyze the development of the concept of water as an economic good and the corresponding strategies of water transnational corporations. This part also focuses on the new values and ideological underpinning of this new water paradigm, which have helped justify and legitimize this new global institutional framework.

I also show how this new water paradigm—in particular the growth of public-private partnerships between transnational corporations, nongovernmental organizations, multilateral organizations, and the state—may well announce a redefinition of the role of the state in the new global arena. This is a redefinition in which the state will have become largely subservient to private global interests, and more generally instrumentalized by transnational corporations, with the help of multilateral organizations and international NGOs. At the same time, the brief case studies of privatization in the Philippines and Bolivia illustrate that this process is not uncontested. In these cases, privatization and sharp price increases led to substantial protests and social unrest. Combined with other adverse economic trends, states were pressured to accommodate some of these social demands, and foreign capital withdrew.

Setting the Stage

Usually underestimated in business studies, environmental issues have become a significant "accelerator" of private sector participation in water management. In our case, environmental issues and problems have laid the ground for the creation of a new water paradigm, which is now being used to justify private sector participation. This section traces this evolution over the past thirty years.

Overview of Water Problems—Underlying Factors and Consequences

The increase in world population, estimated to reach nearly 8 billion by 2020 (WRI 1998), together with industrial development and the practice of water-intensive agriculture, has led to a massive expansion of water resources infrastructure over the past century and to a sevenfold increase in global fresh water withdrawals between 1900 and 1995 (Gleick 1998). With the total amount of fresh water naturally available in fixed supply (World Commission on Dams 2000), the increase in fresh water withdrawals has placed additional pressure on the quantity and quality of available water resources.

The scarcity of water is further reinforced by pollution and contamination from human settlements and industry, which have seriously degraded water quality in many rivers, lakes, and groundwater sources (WRI 1998). Furthermore, the water crisis has been exacerbated by poor urban water management practices and the underpricing of water, which is, as we will see later, considered by many to be the "basic reason for the inappropriate habits of supplying and using water" (Winpenny 1994).

It is now estimated that more than 1 billion people lack access to safe water supply and close to 3 billion are without sanitation. Although 2.4 billion people have gained access to water and an estimated 600 million have gained access to sanitation services over the past two decades (World Commission on Dams 2000), much of these gains have been offset by population growth. Access to water is closely related to income distribution, which has become more unequal in many countries during the last two decades. A survey in Accra, Ghana, found that 55 percent of low-income households had no access to a water source at their residence, compared with 4 percent of wealthy households. Similarly, 60 percent of low-income households shared sanitation facilities with more than ten households, compared with only 2 percent of wealthy households (WRI 1998).

In 1990, an estimated 600 million people in the developing world lived in urban environments, in conditions that constantly threatened their lives and health (Linares and Rosensweig 1999). Pathogenic microorganisms can have a serious impact on human health, causing diseases such as typhoid, gastrointestinal diseases, hepatitis, and cholera. Toxic substances such as heavy metals (mercury and cadmium, for example) and persistent organic substances (such as polychlorinated biphenyls—PCBs, and certain pesticides and herbicides) can acutely or chronically impair the health of humans, animals, and ecosystems. PCBs and other persistent organic substances are known to pass through the marine food chain (IHE 2000). Nutrients, phosphorus, and nitrogen are abundant in municipal sewage, which can lead to overfertilization and algal blooms.

Apart from the devastating impact of lack of safe water access on health, the economic impact is also substantial. The World Commission on Water for the 21st Century has indicated that low income households dependent on water vendors or kiosks on average pay twelve times as much for water as people who obtain water through a private connection. In Karachi, for example, the poor pay 83 times more, while figures of 60 times and 100 times more have been reported from Jakarta and Port-au-Prince, respectively (The World Commission on Water for the 21st Century 1999).

The lack of sanitation coverage and wastewater treatment also has tremendous impact on freshwater ecosystems and the marine environment. A study by the United Nations Environment Program (UNEP) of land-based sources and activities affecting marine, coastal, and associated freshwater environments found that "detailed studies and analysis through the whole [Western and Central African] region, including the landlocked countries, show clearly that sewage constitutes the main source of pollution as a result of land based activities" (UNEP 2000).

Actions to Address the Crisis

Global concern for freshwater problems can be traced to the United Nations' Conference on the Human Environment in Stockholm in 1972. More specifically, water as a key environmental issue emerged during the 1977 United Nations' Water Resource Conference at Mar del Plata. This conference constituted the first real attempt by international organizations to alert the international community to the dangerous overuse of water resources and the growing water scarcity observed in many regions of the world. Not surprisingly, the conference was still part and parcel of the development paradigm, meaning that environmental concerns such as water were mainly perceived as being in opposition to economic development. The concern was that development might jeopardize access to clean water, which was posited in terms of human rights and needs: "all peoples, whatever their stage of development and their social and economic condition, have the right to have access to drinking water in quantities and of a quality equal to their basic needs" (United Nations Water Conference, Mar del Plata 1977, quoted in UNICEF 1995, 5).

Over the past decades, this unfulfilled goal has been recognized at various international conferences and fora. The International Conference on Water and the Environment, held in Dublin in 1992, concluded that the scarcity and misuse of fresh water posed a serious and growing threat to sustainable development and environmental protection. The conference furthermore stated that "concerted action" is required in order to reverse the present trends of overconsumption and pollution. In reaction to the growing crisis, the Dublin Statement set out recommendations for action, based on the following guiding principles:

• fresh water is a finite and vulnerable resource, essential to sustain life, development, and the environment;

• water development and management should be based on a participatory approach, involving users, planners and policymakers at all levels;

• women play a central part in the provision, management, and safeguarding of water; and

• water has an economic value in all its competing uses and should be recognized as an economic good. (These principles will be discussed in more detail next.)

The change in perspective from the 1970s is remarkable. Water has been transformed from a basic human right that is threatened by development to an economic good that is essential for development. The references to participation and women suggest that these moves were not uncontested, though the simple listing of principles undermines any implication that these principles might be contradictory.

This change has been accompanied by increasingly urgent calls to action. Water access has been framed as a global crisis, a challenge to political stability, economic growth, and human security rather than as a local or regional issue. Figure 11.1 may serve as an illustration of the increasingly perceived urgency of the global situation.

Not surprisingly, such maps and figures have led the United Nations in its environmental report *Geo 2000* to note, in quite an alarming tone, that water shortage constitutes a "full-scale emergency," where "the world water cycle seems unlikely to be able to cope with the demands

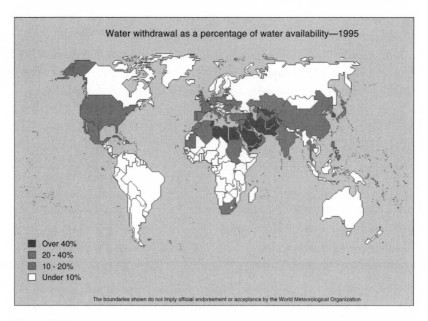

Figure 11.1
Water withdrawals as a percentage of water availability—1995
Source: United Nations Commission on Sustainable Development (1997, figure 11).

that will be made of it in the coming decades. Severe water shortages already hamper development in many parts of the world, and the situation is deteriorating" (UNEP 2000, 362). Such statements can, perhaps, be understood as an effort to raise the profile of the issue to that of other global environmental problems, such as climate change. The effect is to imply that the institutional machinery of global governance is needed to address the problem.

Water as an Economic Good: The Creation of a New Paradigm

The statements at Rio and in Dublin attribute blame for the wasteful and environmentally damaging uses of water on a past failure to recognize the economic value of the resource. Managing water as an economic good, so goes the argument, is therefore an important way of addressing urgency by achieving efficient and equitable use, and encouraging conservation and protection of water resources (Perry et al. 1997).

Principle four of the Dublin Statement, which affirms that "water has an economic value in all its competing uses and should be recognized as an economic good," is also the most controversial principle. Several months after Dublin, in June 1992, the United Nations Conference on Environment and Development in Rio de Janeiro, Brazil, reiterated the Dublin principles, stating that: "integrated water resources management is based on the perception of water as an integral part of the ecosystem, a natural resource and a social and economic good, whose quantity and quality determines the nature of utilization" (Agenda 21, 1992: § 18.6 & §18.8). In particular, this conference also stressed "the implementation of allocation decisions through demand management, pricing mechanisms, and regulatory measures" (Agenda 21, 1992: §18.12 (f)).

In short, both Dublin and Rio represent a fundamentally new way of thinking about water: the allocation, protection, and conservation of freshwater supply can be optimized through market mechanisms and effective management (Finger and Allouche 2002, 25–26). The Dublin principles were sufficiently vague in order to produce consensus, while leaving the operational contents—particularly over who would be doing the management—unspecified. All subsequent conferences, at least until

Johannesburg in 2002, have endorsed this new market-based approach to water management. However, it must be noted, that since the late 1990s, a growing number of international NGOs and trade unions have questioned some aspects of this approach by calling for water to be seen as a "universal public good," rather than as an "economic good" (Barlow 1999; Petrella 2001). More precisely, what these actors question is not the urgency of the situation, nor the idea of participation, but that economics—as opposed to politics—provides the best response to such urgency.

This new approach defined at Rio and Dublin does not in itself emphasize the role of the private sector in managing water. Rather, these conferences created a sense of urgency and then argued that an economic approach—with participation of all concerned actors—is the most appropriate means of addressing such urgency (Finger and Allouche 2002, 28–37). In particular, it is presumed that the internalization of all costs—especially environmental costs—would encourage water conservation (or perhaps more precisely would limit water use), and thus solve environmental problems.

From the outset, the definition of water as an economic good appears to be quite simple, if not simplistic. Says John Briscoe, one of the World Bank's senior water advisors: "like any other good, water has a value to users, who are willing to pay for it. Like any other good, consumers will use water so long as the benefits from use of an additional cubic meter exceed the costs so incurred" (Briscoe 1996, §1). Other authors put the same idea into more technical language when saying:

general principles . . . in assessing the economic value of water and the costs associated with its provision [are]: First, an understanding of the costs involved with the provision of water, both direct and indirect, is key. Second, from the use of water, one can derive a value, which can be affected by the reliability of supply, and by the quality of water. . . . Regardless of the method of estimation, the ideal for the sustainable use of water requires that the values and the costs should balance each other; full cost must equal the sustainable value in use. (Rogers et al. 1998, 5)

The full cost of water is therefore the full economic cost plus any externalities. Environmental externalities are those costs associated with public health and ecosystem maintenance, while economic externalities are all the other costs (such as the costs associated with the impact of an upstream diversion of water). In an ideal situation of sustainable use of water, the full cost must equal the sustainable value in use.

This idea of water being an economic good is indeed becoming increasingly accepted. James Winpenny, a former research fellow at the Overseas Development Institute (ODI), has written an entire book on the benefits of considering water as an economic good. Winpenny's argument is very similar to the one outlined previously: first, water must be recognized as a limited resource (urgency). Not doing so in the past has led, he says, to mismanagement of water, especially by agriculture, industry, and politicians more generally. The solution is therefore to calculate the real cost of water. This will, Winpenny argues, generate additional revenues, which in turn will make it possible to increase water supply coverage, as well as to encourage private sector participation, (meaning investments). Says Winpenny: "low prices depress the profitability of investment in the water sector" (Winpenny 1994, 18). He sees a favorable environment for such a policy in a clear separation between the operator and the regulator, as well as in indicators that provide for real costs, water tariffs, and demand management. The key focus of such "water resources management" must, according to Winpenny, be on the demand side. Winpenny concludes that treating water as an economic good brings a harmony of financial and environmental benefits.

Private Sector Participation

Although not originally on the agenda, private sector participation has become linked to water management via the same narrative of urgency and a need for "better management." As emphasized during the Habitat II Conference for the case of urban management:

The urban environmental crisis will continue to be one of the most pressing problems facing humanity in the twenty-first century. Most of the world's gravest environmental threats to air quality, water quality and availability, waste disposal, and energy consumption are exacerbated by the high density and activity of urban life. Governments acting alone cannot successfully address these challenges—what is needed are partnerships between local governments, the private sector and citizens' groups working together to find solutions. (Habitat II 1996)

Privatization is seen as a way to respond to the sense of urgency, as it introduces commercial principles, professional management, and competition.

The introduction of market pricing using a "full-cost" approach should, according to this new water paradigm, be applied to the water sector so that users realize that water has a cost and that its supply is not unlimited. Higher prices could encourage end-use conservation, as well as lead to new investments in order to increase supply. Market pricing raises issues of access and equity, however. In some countries, such as South Africa, the government ruled that households should receive the first 6,000 liters of water per month for free, after which they would have to pay according to a sliding scale. A full cost approach, where water pricing takes into account social issues, does not in itself contradict the idea that water should be a basic right for all.

Although public water systems could theoretically apply full-cost pricing, there are frequently political constraints to doing so. The new water paradigm closely links full-cost pricing with the introduction of other market mechanisms, particularly private sector participation. Private participation means that private firms can engage in various aspects of the delivery, financing, or ownership of a good. Until recently, most water supply systems were provided by public operators due to three major production characteristics—natural monopoly, strong externalities, and high sunk costs—that give rise to market failure, meaning the inability of an unregulated market to achieve allocative efficiency. It is these features, as well as social and health reasons, that have historically justified the direct provision by the public sector of water supply and sanitation. The logic of private sector participation in water supply and sanitation derives from the conjunction of the twin discourses of water crisis and the inadequacy of the public sector. Privatization is legitimated by the broader neoliberal discourse that links market pricing, deregulation, and private ownership together as the rational, efficient solution to economic problems. Such discourse is promoted by industry associations and international financial institutions.

The New Environmental Governance Structures for Water

Institutions define the rules within which actors operate, thus cementing and legitimizing power relationships. In this section, I show how the private sector and civil society become involved in actively setting up this newly institutionalized water paradigm. The institutions, practices, and ideology of this new paradigm represent a structure of environmental governance, in the broad sense of the term. The term "environmental regime" is often used to mean a rational institutionalized arrangement to solve collective action problems, generally with the state at its center. The new water paradigm, instead, is a mechanism of governance that redefines the power relationships among the involved actors so as to advantage some (particularly, the transnational corporations), as well as to legitimize these relationships by involving civil society actors. In other words, the emerging governance structures for water define a new hegemony in the Gramscian sense.

The link between an economic approach to water resources and private sector participation in the management of the infrastructure has been actively promoted by the World Bank's views and policies. According to a recent study of the 276 "water supply" loans approved between 1990 and November 2002, at least 84 were conditional on the privatization of water utilities <www.icij.org>. Bank officials claim that water projects with private sector participation perform better than those run by local governments. More precisely, water has been redefined by the World Bank as a managerial issue, hence the new term "water resources management." This concept has been introduced in a 1993 World Bank Policy Paper (World Bank 1993), which subsequently served as a guideline for the Bank's ideas on water management at the local, national, and international levels.

According to the Bank, the lack of water pricing and the reliance on government subsidies have been the main factors that have led to the unsustainable use of water resources. As a result, "efficiency in water management must be improved through the greater use of pricing and through greater reliance on decentralization, user participation, privatization and financial autonomy to enhance accountability and improve performance incentives" (World Bank 1993, 40). Water should therefore be considered like any other private good (World Bank 1994, 23). Moreover, it is against this background the World Bank tries to convince people that privatization is not only justified for efficiency reasons, but also beneficial for solving environmental problems. Todd Hanson and Laetitia Oliveira of the International Finance Corporation's (IFC) Technical and Environment Department go as far as to say that "privatizations can make substantial contributions to sustainable development" (Hanson and Oliveira 1997, 34).

This institutionalization of the new water paradigm takes place via new forms of so-called "public-private partnerships." There are basically five such types of partnerships:

• Through a concession contract, a private water operator assumes responsibility for operating existing infrastructure assets for a period of twenty to forty years with a commitment to improve the assets and provide new ones. During this period, the concessionaire is responsible for delivering the service to customers and, throughout the contract, financing the investments involved.

• In a B.O.T. (build, operate, transfer) contract, the operator finances, builds, and operate the new facilities for drinking water production, wastewater, or sewage treatment. The B.O.T. operator bills the public authority for the service. Unlike the concession contract, the end-customer is not the user but the public authority.

• In an affermage contract (or lease contract), the operator manages the service and maintains the facilities for a period of ten to twenty years. The contractor finances the renewal of existing facilities, but is not responsible for financing new infrastructures.

• Operation and maintenance (O&M) contracts grant the operator complete responsibility for the operation of facilities and is paid for the service by the public authority that owns the infrastructure.

• A management support contract gives the operator the responsibility for performing a limited number of tasks with specific objectives, such as reducing leaks, increasing invoicing or recovery rates, and improving the customer relations. The operator is paid by the contract-granting authority according to a schedule of rates that can be adjusted to success in meeting of performance criteria.

A Case Study: Cochabamba, Bolivia

(These two case studies are extracted from Slattery 2003).

The government of Bolivia chose the concession model as the most appropriate form of public-private partnership for the delivery of water and wastewater services in Cochabamba. In 1999, Bolivia awarded a forty-year concession for the Servicio Municipal del Agua Potable y Alcantarillado (SENIAPA) to Aguas del Tunari (a consortium led by the British firm International Water, a subsidiary of the U.S. multinational Bechtel), which was the only bidder for the contract.

The concession was made possible only after the passage of Law 2029—the Drinking Water and Sanitation Law—in October 1999. The law, developed and passed with little stakeholder involvement, created the legal framework to enable private-public partnerships (PPP) in the water sector by enabling private firms to take on the legal responsibility for service provision, and authorizing the privatization of water sources.

Law 2029 not only created the legal basis for PPP in the water sector— it also effectively ended subsidies to the sector. The problems with the Cochabamba concession began when subsidies were eliminated and the price of water was tripled in January 2000. The problem was not only that the utility's existing customers would now have to pay more for water, but that peasants (largely Quechua Indians) in the surrounding arid areas would have to begin paying for water that previously had been theirs free of charge. In a country where 70 percent of the population lives below the poverty line, increases in the price of water can have a serious impact.

The price increases in Cochabamba were partly due to the elimination of subsidies and the need for the tariff to reflect the true economic cost of service, but they were also necessary to pay for the Misicuni scheme— a U.S.$300 million project involving the construction of a dam, tunnel and water purification plants that would boost water supplies to the Cochabamba area.

Opposition to the concession arrangement was formalized in January 2000 when a group of Cochabamba citizens formed "La Coordinadora de Defensa del Agua y de la Vida" (the Coalition in Defense of Water and Life). Through the Coalition, opposition groups rallied and shut the city of Cochabamba down for four days using general strikes and mass mobilization. The culmination of the demonstrations was the signing of the Cochabamba Declaration, a manifesto that called for the reduction of water prices and protection of universal rights to water. This was followed soon after by another demonstration calling for the repeal of the Drinking Water and Sanitation Law, a ban on privatization, repeal of the

Cochabamba concession, and citizen participation in the formation of a new water resources law.

Unfortunately what began as peaceful protests led to rioting and the death of several protesters, and in response the government of Bolivia declared martial law in April 2000. Despite the government's attempts to quell the unrest, opposition to privatization in general and the Cochabamba deal in particular proved too strong. Soon after the April 2000 protests, International Water pulled out of Bolivia and the government of Bolivia repealed its water privatization legislation. Following the departure of International Water, the management of SEMAPA was turned over to its employees and the citizens of Cochabamba, and La Coordinadora has retained an active role in promoting participatory approaches to water management.

Bechtel has sued the government of Bolivia, and is asking to be reimbursed for the approximately U.S.$25 million in investments it has made to date in the Cochabamba system and to prepare for the Misicuni scheme. The suit is currently being heard by the International Center for the Settlement of Investment Disputes (ICSID), part of the World Bank family of organizations.

The concession in Cochabamba was the first of several high profile water deals to fall apart, and the causes for its failure provide some important lessons for policymakers considering public-private partnerships. The issue of tariff increases is always a sensitive one, particularly in countries such as Bolivia where poverty is widespread. However, concession arrangements such as the one in Cochabamba attract substantial private investment for system improvement and expansion while transferring commercial and investment risks to the private operator, which must remunerate itself and earn its return through collection of the tariff. As a result, a concession is only appropriate when the tariff is set to fully recover the cost of operations, maintenance, and depreciation.

In Cochabamba, the problem was twofold. In order to achieve cost recovery and fund the Misicuni scheme, Aguas del Tunari had to raise prices for its existing customers. At the same time, it required those consumers with private wells to shut them down and begin buying their water from the utility. As a result, many consumers who for generations had received free water would now have to begin paying.

The closure of private wells is common practice under PPP arrangements—while critics might see it as a latter day enclosure of the commons, the usual rationale given is that it is in the public interest: the utility can reduce the average price of water for individual consumers by expanding its customer base. In Cochabamba, the private operator had the legal authority to require well closure because Law 2029 granted it the exclusive right to groundwater in the region, as well as exclusivity to provide water services. Many governments around the world have made this transition from private wells to piped service successfully by phasing tariff increases over time and providing consumers with financial assistance aimed at making connection fees affordable.

While the immediate and drastic increases in the water tariff were no doubt largely to blame for the unrest in Cochabamba, many of the problems could, perhaps, have been avoided through early efforts in stakeholder consultation and communication. Law 2029 was created with little input by stakeholders and as a result, the terms and conditions of the concession in Cochabamba were ill-suited to the cultural, political, and economic realities in that area.

Finally, the serious question remains about whether the concession itself was financially viable to begin with. The World Bank had initially encouraged the government of Bolivia to enter into a PPP arrangement. However, after analyzing the proposed investment program, including the Misicuni scheme, and studying consumers' ability and willingness to pay, the Bank concluded that the deal was not financially viable and communicated its opposition to the Bolivian government. The Bank was not the only organization with doubts about the deal's viability. Several of the other consortia that had indicated an interest in bidding on the project decided to withdraw from the competitive process, leaving only Aguas del Tunari to submit a bid. Consequently, public perception of the competitive process was poor and the government arguably did not benefit from the price reductions that typically come from a robust competitive bid. The case suggests that ideological pressures for private governance outweighed even narrow calculations of economic interest. It also points to the success of social forces in contesting privatization.

Manila Water Case

In 1997, the government of the Philippines entered into two PPP arrangements for the provision of water services in Manila. A twenty-five year concession agreement was signed between the Metropolitan Waterworks and Sewerage System (MWSS) and Maynilad Water Services, Inc. (MWSI) for service to approximately 6 million consumers in the Western area of Manila. Maynilad is 59 percent owned by the Philippine-based Lopez family's Benpres Holdings (which also owns Manila Electric Co.) and 40 percent owned by the French water giant, Ondeo.

The agreement was terminated on February 7, 2003, and responsibility for water services reverted to MWSS. Per the terms of the concession agreement, the termination procedure was initiated on December 9, 2002, when Maynilad issued a "Notice of Early Termination" to MWSS, citing MWSS's serious breach of its obligations and giving 60 days' notice.

Although Maynilad cited numerous factors behind its decisions to terminate the contract, including severe droughts caused by the El Niño weather phenomenon, the effects of the 1997 Asian financial crisis, and delays in the completion of a river basin project, the primary reason for termination was MWSS's refusal to implement a rate adjustment that would have enabled Maynilad to recover large foreign exchange losses. Maynilad's bid for the concession, which was mounted just prior to the 1997 Asian financial crisis, was based on the assumption that the peso would depreciate no more than 2 to 3 percent. To compensate for any significant change in exchange rates, the bid included a provision for the recovery of foreign exchange losses throughout the term of the concession via an extraordinary price adjustment mechanism.

Prior to termination, Maynilad's investors had contributed U.S. $143 million in equity and guaranteed a U.S. $165 million bridge loan, in addition to assuming MWSS foreign and domestic debts amounting to U.S. $800 million. During the first five years of operations, Maynilad realized losses of 4 to 5 billion pesos (approximately U.S. $75 to 95 million), with actual annual revenues just enough to service the inherited MWSS debt. When the Asian financial crisis resulted in depreciation of the peso from P26:$1 to P53:$1, Maynilad's debts effectively doubled and the company was no longer able to sustain its operating losses.

The privatization of Manila's water system in 1997 was considered the largest PPP arrangement of its kind to date. The termination is the culmination of six tumultuous years in which Maynilad was faced with protests demanding water connections, opposition to rate increases and rate re-basing, and claims of poor customer service. Despite these difficulties, Maynilad achieved some significant improvements in service, including increasing coverage from 58 percent to 84 percent. However, when compared to the positive experience of Manila Water, the concessionaire serving the eastern zone of Manila, Maynilad's poor performance raises a number of questions about what went wrong. Was MWSS's refusal to increase rates to blame? Was it poor management of the business by Maynilad? Or did extraneous factors such as El Niño and the Asian financial crisis create an environment in which the concession was doomed to failure?

The answer is likely a combination of factors. At the time the concession was awarded, no one—including the Philippine government, its advisors at the International Finance Corporation (IFC), and bidders such as Maynilad—could foresee the effects of the impending Asian financial crisis. As a result, the terms of the concession agreement were drafted such that the impact of ordinary currency fluctuations—on the order of the 2 to 3 percent predicted by Maynilad—would be treated as "pass-throughs" in the tariff, with tariff increases phased in over the life of the concession. This provision significantly reduced the risk to the concessionaire and made the transaction more attractive given the large amount of foreign debt to be assumed from MWSS. While the risk to the concessionaire was reduced, the risk to consumers increased as they would be expected to bear the burden of higher tariffs. If currency fluctuations had been of the magnitude predicted by Maynilad, the corresponding tariff increases would have been gradual and reasonable. Instead, when dramatic devaluations resulting in the need for larger tariffs increased, the regulator was forced to balance Maynilad's financial requirements with consumers' affordability.

In addition to its financial difficulties, Maynilad was unable to achieve operating efficiencies equivalent to those of its rival Manila Water in the eastern zone. By the fourth year of operations, Maynilad's operating expenses were a third higher than the company had projected in its bid, whereas Manila Water's were a quarter lower than it had projected. At the same time, Maynilad's cost per cubic meter to produce and sell water (calculated by dividing revenues by billed daily volume multiplied by 365 days) was P9.06 per cubic meter, as opposed to Manila Water's P4.71 cost per cubic meter. Arguably, the operating environments and therefore the investment requirements in the two concession areas were markedly different, necessitating differing cost structures. However, as one Philippine journalist put it, "It takes Maynilad almost twice as much as the competition to produce and sell a cubic meter of water."

The experience with PPP in Manila's west zone has still not ended. Maynilad and MWSS are in arbitration proceedings to determine exactly how and under what terms the contract will terminate. Meanwhile, Manila Water continues to provide service in the east zone of Manila under its long-term concession.

Legitimizing the New Water Paradigm

Although there are no formal international agreements, rules, or organizations dealing exclusively with water management, one can see that the water sector is increasingly becoming structured along the key principles described before. Not surprisingly, a series of new organizations and alliances have recently emerged, encompassing water TNCs, international financial institutions, water organizations, United Nations agencies, and international NGOs. They have developed networks, at both the international and local levels, which serve the purpose of facilitating as well as legitimizing private sector participation in the water sector.

Indeed, among the key advocates of TNCs in the water sector one will find new "partnerships" involving international financial institutions, water organizations, United Nations agencies, and international NGOs. The aim of such partnerships is to demonstrate that water TNCs, in collaboration with international NGOs and international financial institutions, can indeed fulfil the different functions historically attributed to governments when it comes to water management. Under the auspices of programs such as the "Business Partners for Development Program," one can see emerge initiatives where the private sector works hand-in-hand with NGOs, instead of governments, to provide access to water.

For example, in this particular program, one can find subsidiaries of Vivendi Environment (i.e., Générale des Eaux), which work with the international NGO WaterAid, as well as with the World Bank, in order to provide access to water in urban areas, including to the poor. In other words, one notes the emergence of new institutional partnerships being set up in experimental cities in the South, purportedly to represent the different stakeholders within society.

Moreover, one can find at the international level a new set of structures that firmly establish and legitimize such partnerships. The most obvious example is the Global Water Partnership (GWP). The GWP was officially inaugurated in August 1996 at the Stockholm Water Symposium. Membership in the GWP is open to development agencies, international organizations, NGOs, professional associations, academics, as well as to the private sector. The organization hopes that such diverse partnerships will enable an integrated water resources management policy, as spelled out in the Dublin principles. Government is treated, at best, as an equal partner along with local and international NGOs, as well as with the private sector.

The emerging consensus among international financial institutions, water organizations, UN agencies, and international NGOs, regarding the benefits of introducing private sector participation (PSP) in the water supply and sanitation sector reveals much deeper ideas concerning governance, decentralization, and participation. The private sector, the international financial institutions, and even international NGOs all have a shared interest in diminishing the traditional role of the state in the water sector and have created different forms of alliances to pursue this strategy. The imposition of these forms of governance based on private sector participation clearly illustrates the coming of age of a new way of managing water. It is of course important to understand that this trend is not limited to the water sector, but can also be observed in other infrastructure sectors, such as telecommunications, energy, and transport. Next, the implications of this new water paradigm are discussed, highlighting in particular the relationships between state and transnational corporations.

Business Partners for Development, Global Water Partnership, Global Development Network, and Global Knowledge Partnership, are all examples of partnerships under the UN's "Global Knowledge and Learning Networks." In the normal course of things, the World Bank and International Finance Corporation are frequently also partners in these programmes. In many cases, however, the World Bank acts directly and without even seeking to legitimize its action by involving NGOs or other civil society actors. In January 2001, for instance, Vivendi Water was awarded a EUR 150 million, ten-year renewable lease contract to provide water services for the entire country of Niger, following a World Bank-sponsored international tendering procedure. The World Bank would also provide most of a EUR 35 million investment finance package. Before this date, in November 2000, Vivendi was awarded a U.S. $25 million contract with Nairobi to cover all aspects of water management, including establishment of reservoirs and distribution systems. The tendering process had been conducted by the World Bank in the early 1990s but its implementation was delayed by the Bretton Woods institutions' suspension of aid to Kenya.

As for Bechtel/International Water, the World Bank and International Finance Corporation acted as guarantors to form Manila Water in the Philippines, a consortium between Ayala Corporation, Bechtel Enterprises, United Utilities, and Mitsubishi Corporation. In January 2001 International Water was awarded a thirty-year water concession in Guayaquil, Equador (Ecuador's largest and most populous city) after receiving a guarantee from the World Bank's Multilateral Investment Guarantee Agency (MIGA) worth U.S. $18 million. The guarantee acts as an insurance plan against "the risks of expropriation and war and civil disturbance" and also includes a "performance bond" that "guarantees the company's successful management, expansion, and operation of the water services, against the risk of wrongful call." This is the first time MIGA has issued such a guarantee for a water project.

In May 2000 Suez took over the Cameroon government's 51 percent stake in the state water company "Société Nationale des Eaux du Cameroun" (SNEC) and gained a concession to operate their water supply for a twenty-year period. This followed a joint mission of the IMF and World Bank, which urged the government to accelerate the privatization of SNEC as part of IMF structural adjustment facility for Cameroon. And in 1997, when the World Bank backed and organized the privatization of the water in Manila, Philippines, one-half of the city was given to a consortium led by Suez-Lyonnaise, the other to International Water, a UK/U.S. consortium.

When it comes to private sector participation, the World Bank has quite a simplistic view. For the Bank, private sector participation means, above all, the reduction of the role of the state and privatization; although competition is frequently mentioned as desirable, in the water sector such competition appears to be quite limited (Finger and Allouche 2002, 72–73), often reduced to bidding procedures where firms compete for supplying an entire market. Should monopolies still persist, the Bank considers that the *"transfer to private ownership generally yields efficiency gains"* (World Bank 1994, 53). This last affirmation clearly shows the dogmatic thinking of the Bank, which presumes that public monopolies should be replaced by private monopolies. According to the Bank, the government's role must be strictly limited to regulating such private monopolies by means such as price-caps or rate-of-return regulation. By now, this new kind of thinking seems to have spread and become accepted among a new set of actors, in particular international water institutions, who not only promote this view but legitimize it by involving civil society actors.

The Instrumentalization of the State as a Result of the New Water Paradigm

It is generally assumed that multilateral financial institutions, international NGOs, and (water) TNCs all have an interest in a weakened state. However, some of these actors may also have an interest in a strong state, provided that they can influence its behavior, for example, by influencing the broader institutional framework of which the state is part. Indeed, this is precisely what is currently happening in the water sector. In this section, I make the argument that the state has not become obsolete, but rather has become instrumentalized by TNCs as part of a new institutional framework and discourse, in particular the discourse on "partnerships." The purpose of such instrumentalization for firms is to derive commercial advantages and profits, while simultaneously having such practices legitimized by civil society involvement and NGO participation. This phenomenon is in fact not limited to the water sector. As a matter of fact, this evolution towards state instrumentalization can be observed since 1997, when the World Bank published its report entitled *The State in a Changing World*. In it, the World Bank significantly modified its previously held position, by stating that a "small but strong state" may be better for development overall, as such a state is more likely to attract investment and offer the legal security required for the private sector to operate. And such a small but strong state is precisely what the TNCs want, in the water sector and elsewhere. In this sense, water and its management is merely a paradigmatic case of the future status of the nation-state in a globalized economy, as well as the future power relationships between TNCs and the state. In order to make this argument, I review the key functions of the state, and then demonstrate how these functions are currently made to serve the TNCs as part of the new water paradigm.

The key functions of the state are legislative, the operational provision of services, regulation, and the provision of investment. The state is undergoing profound transformations, and its instrumentalization today can only be understood in this context. Without tracing in detail the history and nature of this transformation, the main drivers have to do with globalization and, more generally, the emergence of powerful nonstate actors. Such actors are not only TNCs, but also multilateral organizations, as well as some international organizations. Many of these actors now have the power to put pressure on the state and in some cases even use it for their own purposes (Finger 2002). Of course, this is a somewhat a dialectical relationship, as the state has significantly contributed to the emergence of the very TNCs that are now instrumentalizing it.

If the nature of these four functions has not changed substantially over time, their relative importance has. Consequently, the service delivery function has somewhat diminished in importance, given the increasing role of the private sector. Also, the investment function has somewhat diminished. Instead, the state has moved from a direct investor to a guarantor of credits and loans, not the least, as we see next, for TNCs. On the other hand, the regulatory function of the state has substantially increased in parallel with growing liberalization and privatization.

Indeed, at least in the developing countries, the nation-state is evolving towards a "regulatory state," which makes its instrumentalization particularly interesting for TNCs. Finally, the legislative function has not changed much over time. However, legislation is increasingly linked to regulation and is becoming of growing interest to the new global actors.

The most significant function to be instrumentalized is service delivery. If one looks at the main TNCs providing water services today, one will find three French companies, Suez, Vivendi, and Bouygues. Their strategy is to seek to provide water services in the most densely populated urban areas, as maintenance and investment cost are much more easily recoverable. Consequently, it appears that many big cities of developing countries are now under some form of management contract with one of these three French TNCs. So far, among the world's biggest agglomerations (more than 11 million inhabitants), only Buenos Aires and Shanghai have a concession contract with a French operator in water supply (respectively, Suez and Vivendi). But many other big cities already have private French operators, namely Santiago de Chile (Suez), La Paz (Suez), Bogota (Suez), Casablanca (Suez), Amman (Suez), Kampala (Suez), Jakarta (Suez), Niamey (Vivendi), Chengdu (Vivendi), Abidjan (Bouygues), and Dakar (Bouygues). The same is true for industrialized countries, such as Paris (Suez/Vivendi), Berlin (Vivendi), Atlanta (Suez), and Indianapolis (Suez). In developing countries, three distinct geographical zones are emerging: large cities for water TNCs, suburban and smaller urban areas for NGOs, and the rest for the state (Bakker 2002). As a result, water TNCs are gaining the most lucrative markets, while the underfunded state is obliged to bear the costs of providing a minimum level of services in the rest of the country.

Water TNCs are also instrumentalizing the state with regards to its other functions. For example, if one looks at the legislative function of the state, one can see that TNCs have used legislative lobbying in order to facilitate their entry in the water sector. Indeed, if one looks at the TNCs' strategy within the European Union, one can see that these TNCs are lobbying national governments as well as European policy makers for stronger environmental and sanitation standards, often with the help of NGOs (Finger and Allouche 2002, 188–191). These strategies are very effective, since local municipalities have not anticipated such higher environmental and health norms and now cannot afford to pay to upgrade their infrastructures. Consequently, they have to turn to TNCs in order to help them manage their water supply and sanitation systems. Indeed, the cost of cleaning water is very high and increasingly requires large and long-term investments that only TNCs can afford, and expertise that only TNCs can provide.

Finally, the state is also instrumentalized by TNCs in its fourth function, as an investor. Although private sector participation takes different forms, it rarely leads to full divestiture. In most cases, water contracts take the form of delegated management, which is the model favored by the World Bank and the TNCs in developing countries. In fact, the instrumentalization of the state takes place with the state as an intermediary between the Bretton Woods institutions and the water TNCs. Indeed, in most developing countries, the water network system is badly maintained due to insufficient funding. The trick here is to get these countries to upgrade their infrastructure without having to pay for it. This is done by means of World Bank and IMF loans contracted by national governments. With these publicly contracted loans, the private water company will then rehabilitate the public infrastructure. Generally, such rehabilitation is outsourced to a building company, which again is a subsidiary of one of the big water TNCs. Of course, such upgrading is generally done selectively, as only lucrative markets are renovated. Once the infrastructure is rehabilitated, delegated management contracts are signed between the local municipality and the water TNCs for a certain period of time. Such a system enables water TNCs to make profits without taking risks, while the government assumes the financial risks by contracting loans from the World Bank or the IMF.

The instrumentalization of the state is not only limited to TNCs, but also occurs with international NGOs and financial institutions. Indeed, international NGOs often partner with TNCs, especially in peripheral urban areas and slums, helping raise awareness and money, while simultaneously providing the necessary legitimation. In other words, while international NGOs do not generally provide water distribution and sanitation services, as the operational aspect is left to the TNCs, they often play a crucial role in financing by securing support from industrial countries and development agencies. The rural areas, however, remain mostly in the hands of the national governments, leading to their neglect. But it is perhaps in the area of regulation where international NGOs are most active. NGOs also play a very useful function for the TNCs, as they often are active in information gathering on issues such as water use patterns or water quality. Indeed, with the state less willing or able to perform these functions, international NGOs sometimes come to replace its regulatory functions to some extent, thus making the state heavily dependent on them.

The Bretton Woods institutions also use and instrumentalize the state for their own purposes. Influenced by neoliberal ideas, the World Bank and the IMF consider that the state should no longer be involved in operational activities and in particular in managing infrastructure services. Forcing this logic, using Structural Adjustment Programs and other forms of conditionality, onto the developing countries, these financial institutions benefit by securing state backing for loans to private corporations. Moreover, the World Bank and IMF strive to limit the state in its regulatory function, while strengthening its function of protecting (private or World Bank) investments. Indeed, only the nation-state can perform this regulatory function and ensure the legal security required for investments to pay off. Moreover, for the World Bank, regulation must be seen in the context of development, where development is equated with promoting private ownership and foreign direct investment.

Here we summarize the five main uses the state has for TNCs, as well as for international NGOs and multilateral institutions (see also Finger 2002).

• First, the state still is an important source of financing, and this is particularly true in the area of infrastructures. Indeed, the state finances many infrastructure projects directly, or takes loans from lending agencies for infrastructure development or rehabilitation. This is important for TNCs, not only because they can make a profit from such infrastructure development and rehabilitation, but also because infrastructure works constitute a privileged entry point to subsequent delegated management contracts that are even more lucrative.

• Second, the state is a guarantor of stability. Such stability has two aspects: on the one hand, TNCs, international NGOs and the Bretton Woods institutions are interested in political and economic stability, as this is a necessary condition for investments to pay off. On the other, they are also interested in legal stability or security. As a matter of fact, the state is the only possible guarantor of legal security, thus making sure that contracts are respected, bills are paid, etc.

• Third, the state is a guarantor of regular revenues. By granting management contracts, lease contracts and especially by granting concessions, the state allows the TNC to collect fees from their customers. This flow of money is generally guaranteed for an extended period of time, as concessions can run up to forty years at times. In other words, the state grants TNCs a (temporary) monopoly. Of course, such monopolies are generally regulated by the state, but TNCs have often learned how to shape such regulation to their advantage, and they frequently renegotiate the contracts to their further advantage once they have been signed, thus taking states and their citizens hostage.

• Fourth, the state is a risk bearer. As said above, TNCs seek to avoid the risks that come with heavy infrastructures. These are mainly the risks of financing these infrastructures without being sure to recover the costs, but also the risks of owning them. In both cases, the state plays a crucial role, inasmuch as it is the state that contracts the World Bank loans. Also, it is the state or a municipality that generally retains ownership of the infrastructure. Should anything go wrong, the TNC can easily pull out, while the state is tied to the infrastructures it can no longer run by itself.

• Finally, and as can be seen most clearly in the case of the European Union, the state is also a vehicle to enforce norms and rules, which have been defined at supranational levels. This of course also applies to structural adjustment conditions, which are imposed upon a state, but have to be enforced within the national borders. Indeed, strong environmental, sanitation or other standards, in the elaboration of which the TNCs have generally participated, are generally first adopted at a supranational level and then enforced by states. This same mechanism also applies to other supranational organizations, such as the World Bank, the IMF, but also the World Trade Organization, the International Organization for Standardization (ISO), and others. In all these cases, TNCs—often with the active support of global NGOs—lobby international organizations to take advantage nationally and locally of these standards and norms.

Conclusions

The emerging governance structures for the water sector demonstrate how the state is increasingly putting its key functions of service delivery, policymaking, and regulation at the disposal of TNCs and other global actors. Water is, in this respect, a particularly good example of such state instrumentalization: indeed, the special and vital characteristics of water make it all the more difficult for governments to withdraw from their responsibilities, especially regarding the management of the scarce resource and related environmental problems, but also concerning health and sanitation standards. Although faced with rising popular resistance against water privatization, the private sector, and especially water TNCs, have managed to establish themselves, thanks to the new water paradigm, as legitimate actors in this sector.

Consequently, and with the active help of the World Bank and international water institutions, water TNCs are increasingly partnering up with the state to become the sole and only provider of water distribution and sanitation services, at least in the most lucrative urban areas. It is TNCs that are best capable of taking advantage of this new situation, as they can act globally and play states against each other when it comes to obtaining the best financing, investment, and regulatory conditions. International NGOs, in turn, play a useful role in helping states define appropriate regulations as well as in legitimizing the new water paradigm, while the Bretton Woods institutions can offer not only the necessary loans, but also help put pressure onto states so that they create the "right partnerships" with the private sector.

The emerging set of arrangements for water supply and sanitation are very different from the formal international regimes such as those for climate change and ozone depletion. Nevertheless, they do represent a structure of global environmental governance. The norms and principles of this governance structure have been laid out in international forums such as Rio and Dublin, and the particulars of water administration in each city are largely determined by a small set of transnational corporations in the water business and by international financial and development agencies, within an ideological framework founded on neoliberal principles. The water governance system is based on market principles of privatization and market pricing (albeit with a heavy dose of government regulation and guarantees) rather than a set of formal international rules and monitoring. While the details obviously vary from case to case, the new water governance paradigm gives rise to a remarkably similar pattern of outcomes across the globe. Typically, broader access to water for some market segments is accompanied by much higher prices and poorer access for rural and poorer population groups. The discourse of stakeholder participation gives some leverage to NGOs to influence and challenge particular schemes; the cases illustrate that the potential to resist is greatest when water regimes are unstable on economic grounds. The increasing, if contested, influence of TNCs in the governance system is part of a shifting balance of power and roles with states and civil society.

References

Agenda 21 (1992). United Nations Conference on Environment and Development, Rio de Janeiro. New York: UN Publications.

Bakker, K. (2002). From archipelago to network: Urbanization and water privatisation in the South. School of Geography and the Environment. *University of Oxford Working Paper Series*. No. 5.

Barlow, M. (1999). *Blue gold: The global water crisis and the commodification of the world's water supply*. San Francisco: International Forum on Globalization (IFG).

Briscoe, J. (1992). Poverty and water supply: How to move forward. *Finance and Development*, 29, 16–19.

Briscoe, J. (1996). *Water as an economic good: The idea and what it means in practice*. Cairo: World Congress of the International Commission on Irrigation and Drainage.

Cosgrove, W. J., & Rijsberman, F. R. (eds.) (2000). *World water vision: Report, making water everybody's business*. London: Earthscan Publications.

Crozier, M. (1963). *Le phénomène bureaucratique*. Paris: Seuil.

Finger, M. (2002). The instrumentalization of the state by transnational corporations: The case of public services. In D. Fuchs & F. Kratochwil (Eds.), *Transformative change and global order: Reflections on theory and practice* (pp. 133–156). Hamburg: LIT Verlag/Palgraeve.

Finger, M., & Allouche, J. (2002). *Water privatisation: Trans-national corporations and the re-regulation of the water industry.* London: Spon Press.

Giddens, A. (1984). *The constitution of society: Outline of the theory of structuration.* Berkeley, CA: University of California Press.

Gleick, P. H. (1998). *The world's water: The biennial report on freshwater resources, 1998–1999,* Washington DC: Island Press.

Gramsci, A. (1971). *Selections from the prison notebooks.* New York: International Publishers.

Gramsci, A. (1995). *Further selections from the prison notebooks.* Minneapolis: University of Minnesota Press.

Habitat II (1996). *Habitat II: Partnership for the urban environment.* United Nations Department of Public Information, <http://www.undp.org/un/habitat/presskit/dpi1790e.htm>.

Hanson, T., & Oliveira, L. (1997). International Finance Corporation. *Environment Matters,* Fall, 32–35.

International Conference on Water and the Environment (1992). *International conference on water and the environment: Development issues for the 21st century,* <http://www.wmo.ch/web/homs/icwedece.html>.

Linares, C., & Rosensweig, F. (1999). Decentralisation in El Salvador. Environmental Health Project USAID, *Activity Report,* 64.

Perry, C. J., Rock, M., & Seckler, D. (1997). Water as an economic good: A solution or a problem? International Irrigation Management Institute, Research Report, No. 14.

Petrella, R. (2001). *The water manifesto: Arguments for a world water contract.* London: Zed Books.

Rogers, P., Bhatia, R., & Huber, A. (1998). Water as a social and economic good: How to put the principle into practice. GWP TAC Background Paper No. 2. Global Water Partnership, Stockholm.

Second World Water Forum (2000). Declaration of the Hague on Water Security in the 21st Century. <http://www.worldwaterforum.org/Minesterial/declaration.html>.

Schwartz, K., Blokland, M., Nigam, A., & Gujja, B. (2001). *Privatisation of water services in developing countries: Implications for the poor and nature.* Unpublished study sponsored jointly by WWF, UNICEF and IHE.

Slattery, K. (2003). *What went wrong? Lessons from Cochabamba, Manila, Buenos Aires, and Atlanta.* Reason Public Policy Institute, Annual Privatization Report, <www.rppi.org/apr2003/whatwentwrong.html> and <www.ip3.org/publication2003_002.htm>.

United Nations Children's Fund (1995). *UNICEF strategies in water and environmental sanitation.* New York: UNICEF (Document E/ICEF/1995/17, issued 13 April).

UNCSD (2000). *Progress made in providing safe water supply and sanitation for all during the 1990s, Report of the Secretary-General,* Geneva. Eight session. 24 April–5 May 2000 (E/CN.17/2000/1).

UNEP (2000). *Global environmental outlook 2000.* London: Earthscan Publications.

Winpenny, J. (1994). *Managing water as an economic resource.* London: Routledge.

World Bank (1993). *Water resources management.* A World Bank Policy Paper. International Bank for Reconstruction and Development. Washington, D.C.: The World Bank.

World Bank (1994). *World Development Report 1994: Investing in infrastructure.* New York: Oxford University Press.

World Bank (1997). *World Development Report 1997: The state in a changing world.* New York: Oxford University Press.

World Commission on Dams (2000). *Dams and development: A new framework for decision-making.* London: Earthscan.

The World Commission on Water for the 21st Century (1999). *World water vision.* Geneva: UNDP.

WRI (1998). *World resources, 1998–99.* Washington, DC: World Resources Institute.

12

Challenging the Global Environmental Governance of Toxics: Social Movement Agency and Global Civil Society

Lucy H. Ford

The development of global governance is part of the evolution of human efforts to organize life on the planet, and that process will always be ongoing.
—Commission for Global Governance 1995, xvi

Introduction

As other chapters in this volume clearly illustrate, civil society pressure upon firms to improve their environmental performance has increased enormously over the last twenty years. From positions of confrontation and antagonism, there have been moves on the part of both business and civil society organizations to form partnerships to develop solutions to environmental problems. Businesses recognize the environmental credibility and expertise they gain from association with NGOs, and NGOs recognise that working with business provides them a channel of access to actors that are in many ways more powerful than the state actors who have traditionally been the target of their campaigning. In this context there is a need for some conceptual clarity about the difference between the NGOs that are increasingly being afforded a role in global governance and other social movements whose agendas and modes of organizing have largely been excluded, and about the ongoing role of conflict in relations between business and other civil society actors. In order to understand the nature of relationships between business, civil society and the state around questions of environmental governance, we have to look at the tensions and dilemmas within civil society about whether and how to engage in constructing governance arrangements at the global level. By using insights from the work of Gramsci, it is possible to see how

these relationships are both contested and in flux, amid competing understandings about what is to be gained through engagement. This approach
helps us to get beyond static understandings of civil society, and to
identify the possibility of new constellations of interest in the arena that
provides the case study for this chapter, toxic wastes.

Within orthodox International Relations (IR), global governance is
widely regarded as the solution to problems perceived to be global, such
as environmental degradation. As part of these efforts to "organize life
on the planet," corporate civil society actors have been quick to claim
they are the vanguards of sustainable development and have been lobbying hard to absolve themselves from environmental regulation. Social
movements, on the other hand, have been all but invisible, rarely seen
and never heard at the high table of international economic and political relations. Only recently has the more established end of the spectrum
of social movements—that of NGOs—been accorded a space for
engagement with global environmental governance through the sphere
of global civil society, which has been widely portrayed as a democra-
tising force. However, this orthodox, liberal conceptualization eschews
an analysis of power relations, both inside the sphere of civil society
itself, and between global civil society, the interstate system and the
global market. It does not problematise the conflict of interest between
business and social movement elements of global civil society. Moreover,
it tends to conflate social movements with NGOs. Much has been written
about social movements in global politics, which need not be reviewed
here (see, for example Eschle and Stammers 2001; Keck and Sikkink
1998; Waterman 2001; Willetts 1996). This chapter seeks to theorise the
significance of global civil society as an actor within global environmental governance, focusing particularly on social movements. These are
broadly conceived as lying on a spectrum ranging from grassroots movements to transnational NGOs. This is not to say that social movements
form a homogeneous group. Indeed, their aims and methods are as
diverse as the movements themselves. However, the focus here will be on
movements engaged broadly in working for what they perceive to be
progressive social and ecological change.

In line with the neo-Gramscian perspective outlined in chapter 3,
global governance—including that of the environment—is viewed here

as embedded in the neoliberal global political economy, which is hegemonic in the neo-Gramscian sense that dominant power relations are maintained by consent as well as coercion (Cox 1981, 137). Hegemony is maintained through the discourse and practices of global environmental governance, as it is articulated and structured by global environmental institutions (for example CGG 1995). The chapter begins by briefly tracing the evolving discourse of global environmental governance, which appears to be marked in particular by the inclusion of global civil society and thereby ostensibly social movements. It then provides an illustration of social movements campaigning against toxic waste. It examines the relationship between thinking globally and acting locally, and goes on to argue that all activism must have an analysis of the global, even though actual activism necessarily takes place in a particular locality.

What is global about global social movement activism is the movements' challenge to hegemonic forms of environmental governance. *Global* is understood here as a causal category rather than a spatial term, thus avoiding the conflation of global with transnational or international, as orthodox analysis often does. Unlike the orthodox spatial definition of global environmental degradation, Julian Saurin has argued that environmental degradation is global because it is caused by social, economic and political relationships and structures that are global (Saurin 1993). The emphasis here is on the international political economy of industrial production rather than the transboundary distribution of, for example, greenhouse gases and sea level rise. The chapter goes on to reconsider the sphere of global civil society in a neo-Gramscian framework as the site for simultaneously challenging and maintaining hegemony, and asks to what extent global social movement agency constitutes a counterhegemonic challenge to global environmental governance.

Global Environmental Governance

The orthodox response to environmental degradation has been of an institutional nature, focusing on international cooperation and the ordering and management of the interstate system. This is reflected clearly in the literature on international regime theory (Krasner 1983; Young 1989), which has been widely criticised in ways that need not be repeated here (but see for example Kütting 1997; Paterson 2001 and chapter 2 of this volume). The point for the present purposes is that social movements are absent from conventional analysis, including regime theory. International regimes are the stuff of states, interstate institutions, and certain privileged nonstate actors who provide experts in the epistemic communities that form around particular issues (Haas 1990). But while NGOs may contribute to epistemic communities, the epistemic communities themselves do not embrace the concerns of the more radical NGOs and grassroots movements, which have often emerged as a result of direct experience of environmental problems. Simply put, epistemic communities are part of the problem-solving community, whereas many civil society groups seek to challenge the very premises of an incremental and technical problem-solving approach.

There is much continuity between the international regimes approach of the 1970s and 1980s and the more recent development of a discourse of global governance. The quotation from the Commission on Global Governance (CGG) at the start of this chapter exemplifies an orthodox discourse, which portrays global governance as a natural quest for planetary order. Within this discourse, there is a consensus that global environmental cooperation and management is crucial to dealing with *global* environmental degradation, which in traditional parlance is understood in spatial terms as those issues that affect more than one country, and potentially the whole globe (Porter and Brown 1991).

The orthodox discourse of global environmental governance, however, does not appear to differ much from the old tales and practices of international regimes. Both approaches can be seen as processes of institutionalization that stabilize and perpetuate a particular order (Cox 1981, 136). Richard Ashley argues that a discourse of governance, as such, is about the imposition of international purpose, which centers on the "production and objectification of enduring structures that . . . lend to global life an effect of continuity, of a direction, and of a unified collective end beyond political questioning" (Ashley 1993, 254).

What Ashley calls a "discourse of continuity" constitutes both a temporal and spatial enclosure that thereby forecloses the possibility of change. The boundaries of the discourse are never questioned and the limited scope of a problem-solving approach precludes an understanding of environmental degradation as embedded in the wider global political economy and deeper social relations than merely those of states and experts. In concrete terms, this can be seen in the relatively weak outcomes of "soft" environmental negotiating: for example the shallow outcomes of the 1992 UN Conference on Environment and Development (UNCED), the focus on voluntary arrangements such as the Global Compact and ISO 14000 rather than legally binding measures to enforce corporate accountability, and most recently the widely perceived failure of the 2002 World Summit on Sustainable Development (WSSD) in Johannesburg to fulfil its mandate.

However, there is one key difference between international regimes and global governance of relevance to this chapter, and that is the explicit reference by the latter discourse to the growing importance of global civil society (Young 1997, 2). The CGG Report *Our Global Neighborhood* defines global governance as "intergovernmental relationships, which now also involve NGOs, citizens' movements, multinational corporations, and the global capital market" (CGG 1995, 3). According to this view, what appears to be global about global governance is the inclusion of the sphere of global civil society, which includes a wide range of actors who do not necessarily have shared interests. As we shall see later, the sphere of civil society is a contested space.

The history of this shift from an international regime approach to a global environmental governance approach can be charted in the history of international negotiations. Key to this is the language of participation that emerged out of the Brundtland Report, which provided the groundwork for the UNCED conference and which at first sight appears to present a radical break from the mere involvement of epistemic communities. At UNCED, *Agenda 21* was adopted as a comprehensive blueprint for global action in key areas of sustainable development, including the role of various sectors of society: women, indigenous peoples, NGOs, workers, trade unions, business and industry, and the scientific and technological community (United Nations 1992). In short, almost everyone had become a stakeholder in global environmental governance.

While, for the first time ever, the world's population was being called upon to participate in the saving of the planet, this democratization of environmental governance has not actually been extended to the negotiating table. Further, global civil society, the space where these many voices gather, is a fairly exclusive club that often excludes its grassroots members. Also, this vision of global civil society explicitly includes the corporate sector in its realm alongside NGOs, portraying them as equal stakeholders, eschewing an analysis of power relations within the sphere itself.

The focus of this chapter is on social movement agency in global environmental governance, in particular through global civil society. However, it should be mentioned that quite apart from these manifestations of public global environmental governance there is also a privatization of global environmental governance taking place (see Clapp, this volume). The private sphere of business and enterprise, while participating in civil society dialogues and publicly greening its image, has privately been attempting—and succeeding—in absolving itself from international regulation. In fact, the corporate sector has been instrumental in the promotion of a very narrow conceptualization of sustainable development as sustainable growth, both in the run-up to Rio and since.

During the UNCED conference, Maurice Strong sought to bring business on board and the Business Council for Sustainable Development was born (BCSD, later renamed WBCSD). Curiously, however, business received little attention during the official negotiations at UNCED. Agenda 21 only mentions corporations in order to emphasize their role in sustainable development, but does not acknowledge the need for business to be regulated (Thomas 1994, 19). More fundamentally, at the same time as UNCED was being held, UN reforms were underway that axed the UN Center on Transnational Corporations (UNCTC) (Thomas 1994, 19). Attempts by the UNCTC to include recommendations for corporate accountability were successfully undermined by corporate lobbying (Chatterjee and Finger 1994, 117). The controversy over the lack of provisions in Agenda 21 regarding corporations was further enhanced by the fact that major corporations were found to be contributing to the funding of UNCED itself (Doran 1993, 57).

Together with the International Chamber of Commerce (ICC), among others, business has been working hard to "greenwash" its corporate image. As far back as 1984 UNEP and ICC organised the World Industry Conference on Environmental Management (WICEM), which three years prior to the Brundtland Report's promotion of sustainable development, was discussing the "the state of the art of achieving economic growth and sound environmental management" (Trisoglio and Kate 1991, Preface). The position was a distinctly corporatist one. At WICEM it was recommended that industry should become more strongly involved in formulating environmental policy in general, as well as in formulating national environmental regulatory frameworks (Trisoglio and Kate 1991, 22). By 1991, in the run up to UNCED, WICEM II was clearly carving out the niche for industry:

Industry has to help define sustainable development and play a full role in the UNCED to be held in Brazil in 1992. Business leaders are already demonstrating that more sustainable behaviour is in the self-interest of industry, and that many opportunities accompany the course of change. The growing role and international scope of industry confer upon it growing responsibilities to play a role in the developmental aspects of sustainability, and to propose innovative ways for industry to contribute to global issues such as poverty. (Trisoglio and Kate 1991, 30)

As part of this quest, WICEM II further called on business and industry to foster harmonious relations with local communities in order to gain their confidence and to become better integrated into the community and wider society (Trisoglio and Kate 1991, 30–32). The result of WICEM II was *The Business Charter for Sustainable Development: Principles for Environmental Management*, adopted in 1990 and first published in 1991 (Grubb et al. 1993, 11). This states, for example, that:

economic growth provides the conditions in which protection of the environment can best be achieved, and environmental protection . . . is necessary to achieve growth that is sustainable . . . In turn, versatile, dynamic, responsive and profitable businesses are required as the *driving force* for sustainable *economic* development and for providing managerial, technical and financial resources to contribute to the resolution of environmental challenges. Market economies, characterized by entrepreneurial initiatives, are essential to achieve this . . . making market forces work in this way to protect and improve the quality of the environment—with the help of standards such as ISO 14000, and judicious use of economic instruments in a harmonious regulatory framework—is an ongoing challenge that the world faces in entering the 21st century. (International Chamber of Commerce; ICC 1991, emphasis in original)

The growing role of business in global environmental governance was most recently manifest at the 2002 Johannesburg Summit on Sustainable Development, where business was visibly dominant and succeeded in keeping corporate accountability off the agenda. The arena of global civil society, where business and social movements confront each other, appears heavily skewed in favor of business. In the neo-Gramscian perspective, this can be seen as an entrenchment and legitimation of hegemonic global environmental governance through the sphere of global civil society. However, let us first turn to an analysis of agency as a potentially radical concept, in order to conceptualise social movement challenges to orthodox global environmental governance.

Social Movement Agency

Part of the explanation for the absence and invisibility of social movements in most accounts of world politics lies in the belief that they are simply not the subject matter of interstate relations, that they have no ontological status. As R. B. J. Walker has reminded us, "judged from the regal heights of statecraft, social movements are but mosquitoes on the evening breeze, irritants to those who claim maturity and legitimacy at the centers of political life" (Walker 1994, 669).

Furthermore, deeper epistemological structures are involved in obscuring the role of social movements as actors in world politics as well as inhibiting an understanding of the role of generic agency in analyzing resistance and social change. In IR, as in other disciplines, questions about agents and structures and the relationship between the two have sparked an ongoing debate, referred to as the agent-structure problem. The focus of this debate generally remains with states (actors) and the interstate system (structure). Orthodox IR privileges the "agency" of states within the state system, understood as state "actors" and their observable "actions." While the orthodox framework has traditionally also marginalized social movements, the point is not merely to "bring social movements in," but to focus on radical agency and its potential to challenge to global hegemony, in this particular instance on the part of social movements. This is not to say that there are necessarily direct causal links between social movements, resistance, and change. Resistance and change are mediated through the relationships of power between actors and structures. Following from the earlier definition of "global," where global refers to "a space of causal power," agency is thus also global (Maclean 1999b, 182, emphasis in original). If social movements themselves fail to recognize their own location within global hegemony they may end up reproducing global hegemony rather than challenging it. Maclean warns that:

These contemporary groups may be performing—unintentionally—a crucial historical requirement for the completion of the establishment of the first fully global (as distinct from international) hegemonic formation. A new transnational, or non-national bloc of subordinated but informed consent, complementary to, but apparently separated from the transnational, or non-national, managerial class (Maclean 1999a, 4).

Although it is important to make social movements visible as actors in international relations, it is not sufficient for them merely to gain access to the negotiating table. The discourse within which global environmental governance operates is itself a site of contestation. This is particularly visible in global environmental governance where environmental problems are usually seen as challenges capable of technocratic solution and where the wider social, political, and economic structures that contribute to environmental degradation are not usually contested.

Social movements campaigning against environmental degradation include a diversity of types, employing a range of tactics and strategies. The more institutionalized NGOs that actively engage with global environmental governance tend to work largely within the framework of the technocratic problem-solving approach. By adopting the dominant technical and economic language of current negotiations, without presenting a more radical challenge to corporate practices and economic structures, NGOs can gain a measure of access, legitimacy, and influence. The price to be paid is the risk of cooptation and contributing to the reproduction of existing mechanisms of environmental governance; Gramsci used the term "passive revolution" to refer to this process of modest concessions by dominant groups that broaden the hegemonic coalition to sustain legitimacy, and thus maintain the essentials of hegemony. Other more radical movements reject engagement, pointing out the need for deeper structural change. They seek to challenge the very edifice of technical-rational management, which, they claim, is not capable of resolving environmental problems caused by the irrationalities and ecologically and socially contradictory nature of production, consumption, and distribution within advanced capitalism. Their resistance is an attempt at seizing radical, counterhegemonic agency. However, this strategy risks marginalization, as they remain outside the realm of discourse and their voices, more often than not, go unheard (Bourdieu, 1977, 168). Gramsci appreciated the need for a synthesis of these strategies when he argued that a successful political force "moves on the terrain of effective reality, but does so in order to dominate and transcend it" (p. 172). As I argue next, the challenge may lie in a reconciliation between these diverse ends of the spectrum, in an attempt to challenge the boundaries of discourse and hence practice.

Social Movements Campaigning against Toxic Waste

The issue of toxic waste serves as an excellent example of a growing environmental problem that has given rise to a wide variety of social movements. Increasing volumes of toxic waste and the ensuing waste trade are an inherent, if contradictory, feature of the present global political economy, reflecting patterns of production and consumption that maintain and reproduce global hegemony. Countless social movements, from grassroots community groups to large NGOs such as Greenpeace, are campaigning to expose the damaging nature of toxic waste.

The struggle against toxic waste is not a discrete issue, but one that is fundamentally embedded in the global political economy and one that emphasises unequal social relations of power, throwing up questions of race, class, and gender. Thus, for example, the environmental justice movements that have sprung up to challenge the dumping of waste are often made up predominantly of women of disadvantaged social groups that are disproportionately exposed to hazardous waste. Issues of democracy and accountability come to the forefront during their struggles, as they find they are fighting against bureaucratic hierarchies that tend to side with industry (Gibbs and CCHW 1995). Activists come up against a scientific elite who are defending a particular discourse and rationality that tends to discredit any tacit and lay understandings of the environmental and health consequences of toxic waste (Brown and Ferguson 1995; Brown and Masterson-Allen 1994; Krauss 1993). Furthermore, the movements challenging the toxic waste trade highlight the distributional issues between countries and communities across the globe, and are part of a growing, diverse constellation of global counterhegemonic movements (Hallowes 1993; Puckett 1999). For these groups, the struggle against toxic waste entails a struggle against expert knowledge, especially scientific knowledge as it is constructed and used by industry. Science is not a neutral tool, but is capable of being turned to the use of states to justify inaction in the face of "scientific uncertainty" or for the protection of corporate interests.

One of the first incidents exposing the devastating effects on human health resulting from the dumping of toxic waste was Love Canal. The town suffered decades of highly toxic waste dumping by a chemical subsidiary of Occidental Petroleum. Construction of a school and hundreds of homes on the site took place with the full knowledge of public officials (Gibbs and CCHW 1995). In the words of Lois Gibbs, who initiated the grassroots community movement, the history of Love Canal was one of

cover-ups, lies, and deception; data manipulation by corporations and government as well as fraudulent claims and faked studies . . . It's a story of money and power; of how corporations influence government actions and how this collusion affects the public (Gibbs and CCHW 1995, 1).

Such practices triggered the politicization of waste, exposing the class, race, and gender aspects of waste pollution, particularly regarding the siting of landfills. On the one hand, protests such as Love Canal have triggered a grassroots environmental justice movement, which challenges mainstream environmentalism by questioning prevailing models of industrial development and the interests that benefit from it. On the other hand, the more mainstream NGOs do not necessarily campaign against the underlying structures and processes of power, but for the amelioration of the environment, or the greening of government and business. To be effective, both ends of this spectrum must be reflexive and conscious about their agency, which requires an analysis of their position within global hegemony, their sources of leverage, and the constraints they face.

From NIMBY to NIABY—A Global Movement?

Groups campaigning against toxic waste dumping are often engaged in the specific targeting of a firm or government department. However, their concerns do not thereby remain of a narrow, single-issue type. What started off as a NIMBY (Not-in-my-backyard) campaign fostered consciousness about the toxic organization of society and a deepening of understanding of related issues on a more systemic basis, leading to what some have called a "radical environmental populism" (Szasz 1994, 6). The ideological shift from calls for NIMBY to calls for NIABY (Not in Anybody's Backyard) (Brown and Masterson-Allen 1994, 272) presents a concrete recognition of the social, political, and economic structures of production and consumption across the globe that are contributing to the toxic waste crisis. This is going one step further than mainstream environmental groups, who tend to call for the proper disposal or reduction at source without challenging the underlying model of perpetual growth. The issue of toxic waste, in a concrete way, politicises issues that were previously considered economic (industrial processes, landfills) and private (individual family health), as well as politicizing previously marginalized citizens (Krauss 1993).

As pointed out above, social movements must be aware of their position in global hegemony. John Maclean (1999a) has argued that

social movements and TNGOs (transnational non-governmental organisations) that profess emancipatory aims (and do so sincerely) are bound to misrecognize their own relationships with the deep structure of hegemony in the late-modern world, unless their analysis of agency is articulated through the context of globalisation (p. 3).

The activism around waste highlights dynamics between the local and the global within the environmental movement as well as the global political economy. In particular, it demonstrates a "paradox" relating to activity in developed and developing countries, as seen in the initial NIMBY movement in the North, which contributed indirectly to the dumping of toxic waste in poorer countries, giving rise in turn to environmental justice movements in these countries. Another paradox relates to the emancipatory potential of movements. The initial NIMBY movement, while misguided on a global scale, nevertheless nurtured valid and valuable forms of political emancipation. This self-creative aspect of social movement activism needs to be a part of the reconstruction of radical agency. Radical social movement agency cannot solely be equated with revolutionary agency and juxtaposed to reformist agency. There is a spectrum of social movements with emancipatory potential, however latent and diverse. Gramsci's concept of a "war of position" reflects the need to develop complex, multilayered strategies based on a sophisticated analysis of existing social forces and structures.

Activism around toxic waste is mostly a reaction to concrete local concerns rather than triggered by orthodox "global" concerns such as ozone depletion or global warming (Brown and Ferguson 1995). For these reasons, the environmental justice movements that campaign against toxic waste present a radical challenge to mainstream environmentalism. On the issues of class and gender, for example, the politicization of working-class women through the issue of toxic waste has instilled a suspicion of mainstream environmental organizations that fail to analyze the underlying issues of inequality in access to power and influence (Krauss 1993, 248; Seager 1993). The "global" aspect of grassroots movements forming a decentralized environmental justice movement thus lies in their challenge to global hegemonic structures of production and consumption, unlike mainstream environmental movements that campaign about "global" issues. More and more, movements are addressing issues explicitly in their global context. However, as noted previously, the "global" is not a spatial but a causal category. It is thus possible to think globally and act locally simultaneously. Ultimately, movements that seek to challenge global hegemony have to adopt a holistic approach, which some global social movements are doing by targeting the interconnectedness of environmental and social problems within the global political economy. However, this analysis needs to be reconsidered in the light of global civil society.

Global Civil Society: Democratizing Force for Global Governance or Challenge to Global Hegemony?

Global civil society is usually understood as a sphere of voluntary societal association that is located "above the individual and below the state but also across state boundaries, where people voluntarily organize themselves to pursue various aims" (Wapner 1997, 66). It is interpreted as a space for dialogue between the institutions of global governance—such as the WTO or the UN—and societal actors ranging from NGOs to business and industry lobbyists. In this view, global civil society represents a liberal democratic space that complements the state-system and thus enhances global governance (CGG 1995; Lipschutz 1996).

The representation of NGOs alongside corporate actors corresponds very much with the liberal description in *Our Global Neighbourhood*. In this view, there is little acknowledgment that this is a potential site for conflicting interests. Rather, it is portrayed as a space of "civility." The emancipatory role of business is expressly pronounced in the passage below by Paul Wapner (1997, 77):

Through business interaction, people begin to care about others and to establish modes of conduct that orient their activities. The business of business, in other words, spills over into the broader social domain that contributes to social sensibility; it is thus partly responsible for the "civility" of global civil society.

Wapner invokes the notion of global civil society as a sphere that is "essential to the unity and coherence of hegemonic social orders" (1997, 77). It is unclear, however, how global civil society, thus understood, could also represent a site for challenge. It's counter-hegemonic potential seems to be lost in a "civility," driven by the very corporate actors that social movements see as part of the problem. The neo-Gramscian perspective, however, allows for global civil society to be seen as the terrain for legitimizing as well as challenging global governance. Global civil society, in this conception, has a dual existence, as part of the extended state where it is key to achieving hegemonic order, and as a semi-autonomous, contested space outside the state and the market economy. The notion that global civil society is an autonomous, self-creative, potentially counterhegemonic force is further advanced by Stephen Gill, who argues it is the breakdown of the Cold War and end of interbloc and inter-imperial rivalries that "open up new potential for counterhegemonic and progressive forces to be able to make transnational links, and thereby to *insert themselves* in a more differentiated, multilateral world order" (Gill 1991, 311, emphasis added). This view of global civil society explicitly recognises it as a contested political space.

As illustrated above, social movements must comprehend their position within this hegemonic constellation, and recognise that not only are they engaged in political contestation with other, more powerful, actors such as states and firms, but that there are structural and discursive forces at play, of which the very framework of global civil society is itself a critical part. Participation by social movements in civil society, therefore, involves a fine balance between challenging and reproducing the existing economic and social order. With this awareness, social movements can be creative, emancipatory forces for social change. In particular, their insistence on transparency and accountability of global institutions, their tactics of selectively engaging with and sometimes bypassing states and interstate institutions, as well as their loose coalitions of diverse NGOs, some more professional and some grassroots, may actually contribute to positive, emancipatory cultural and social change by introducing new ways of organizing and interacting. Social movements as an element of the global political economy cannot be ignored, in so far as they are instrumental in getting issues on the agenda, even if the direction that agenda takes remains contested.

Some social movements, particularly the more established NGOs, seek to influence the global agenda by engaging with the institutions of global governance (Williams and Ford 1999). Environmental movements, as well as other actors such as labor organizations or consumer groups, have challenged institutions such as the WTO to increase NGO participation. Such participation remains for the most part informal. While these institutions show willing to address calls for participation and transparency, paying lip service to the rhetoric of *Agenda 21* or *Our Global Neighborhood*, the arrangement also enhances their image while not necessarily altering the terms of the debate. This is not to undermine the vast efforts made by NGOs within the UN system and other global institutions. However, the question of cooptation remains. Even though some NGOs may critique orthodox approaches to trade, environment, and sustainable development, for example, it is questionable to what extent orthodoxy is up for negotiation within fora that are designed from the start by dominant institutions (Ford 1999; Williams and Ford 1999).

In the case of toxic waste, the Secretariat of the *Convention on the Control of Transboundary Movements of Hazardous Wastes and Their Disposal* (Basel Convention) allows NGOs to observe its various meetings. The back row is always taken up by a variety of members of global civil society ranging from Greenpeace to CEFIC (The European chemical industry lobby). The Basel Convention may be an exceptional case, in that over the years NGOs such as Greenpeace have been instrumental in collecting data about the toxic waste trade, and have been at the forefront, together with the developing countries, of campaigning for a ban on the trade in waste from North to South. In this sense, the overall shaping of the Basel Convention has been directly influenced by a major environmental NGO (Clapp 1994, 510). However, at the Conferences of Parties and technical working group meetings, the agenda is already set by the Secretariat and the member states. Although NGOs are welcome to raise their hands during the meetings and make suggestions and objections alongside member states, their power to challenge the agenda per se is very limited. This would suggest that if they wanted to make any substantial suggestions to the draft documents before the actual meetings, they would still have to resort to lobbying individual member states beforehand. They can monitor and contribute to the open meetings that they are invited to, but the agenda setting and preparatory work takes place at the national level, in smaller working groups or within the secretariat itself.

NGOs are thus marginalized within the international negotiating process. In other institutions such as the WTO, they are excluded altogether from the negotiations. Global civil society is, therefore, not necessarily an emancipatory force, as its agency is enclosed in the institutions of global governance, and the NGOs that enter that realm risk cooptation; real decision-making power remains within state-controlled organizations such as the WTO, or World Bank, which pay more attention to the voices of business and industry, or in business organisations such as the World Business Council for Sustainable Development (Chatterjee and Finger 1994, 151ff; Ford 1999, 70).

As seen previously, many movements reject dominant forms of global environmental governance. However, to what extent do they constitute a counterhegemonic force? According to the neo-Gramscian analysis of Robert Cox, a transformation in world order would require "fundamental change in social relations and in the national political orders, which correspond to national structures of social relations" (Cox 1996, 140). For Cox, this project cannot take place within global institutions, as they are capable of "absorbing counterhegemonic ideas" (1996, 138). This line of argument complements the view that global civil society can act as a vehicle of cooptation and mechanism of hegemony. Drawing on Gramsci, Cox sees overtures to NGOs as a concession to "subordinate classes in return for acquiescence in bourgeois leadership, concessions which could lead ultimately to forms of social democracy which preserve capitalism while making it more acceptable to workers and the petty bourgeois" (Cox 1996, 126). The discourse of global civil society as a democratizing force in global governance could equally be seen as part of a strategy for "assimilating and domesticating potentially dangerous ideas by adjusting them to the policies of the dominant coalition and can thereby obstruct the formation of (class-based) organized opposition to established social and political power" (1996: 130). Thus the agency of global civil society is not necessarily radical and emancipatory. In the case of toxic waste, mainstream environmental movements within global civil society are not calling for any radical changes to social structures, but for cleaner and more efficient production and disposal.

Cox argues that counterhegemonic forces need to return to the national context. Although activism can necessarily only take place in a geographical place, that is, a locale, the transnational links are invaluable in terms of creating a shared understanding of global hegemony, however differentially it is manifested across the globe. Although ultimately social movements will return to their individual localities, they are building bridges of mutual awareness of the struggles going on around the globe.

In the spirit of this growing collective consciousness, social movements may need to adopt a symbiotic strategy of pragmatic engagement as well as resistance. An example could be seen in the case of the campaign against toxic waste. During the negotiations of the Basel Convention and in later negotiations, alliances were formed between NGOs and delegates from the Group of seventy-seven developing countries in attempts to push for a ban on the waste trade between developed and developing countries (Clapp 1994, 510). The International Toxic Waste Action Network and later the Basel Action Network (BAN) are examples of networks of transnational movements campaigning against toxics and the toxic waste trade. BAN actively promotes the implementation of the Basel Convention and the immediate ratification of the ban on toxic trade between OECD and non-OECD countries. Further, it acts as a clearing-house for up-to-date information on global toxics and the toxic waste trade, as well as a point of contact for a myriad of grassroots movements across the globe. Members of BAN, as well as Greenpeace, attend the COPs on a regular basis, thus taking on a dual role, engaging with the international policy process as well as resisting on the ground (Puckett 1999, 31). Jim Puckett (1999, 31), director of BAN, has explained:

[W]e must resolve to move beyond acting solely locally, but must simultaneously learn to become active in the rarified air of global politics, and the national policies that shape them. Where we have no access to these cloisteres of bureaucrats we must relentlessly insist on active representation inside.

BAN thus differs from other NGOs that actively engage in the international policy process, such as WWF or the World Conservation Union, that are highly institutionalised and have arguably lost touch with the grassroots of the environmental movement. BAN is a global social movement in the sense that it appreciates and challenges the economic, political, and social structures of the global political economy that are producing the global toxics crisis (BAN Policy Principles, <http://www.ban.org>). While the Basel Convention in itself is part of hegemonic global environmental governance, a ban on toxic waste moving from developed to developing countries may nevertheless be an important incremental step towards progressive social change. BAN is thus engaging in a "war of position" by working within current structures for stronger controls on toxic wastes, while retaining an outside, grassroots critique as part of a longer term, strategic movement for change.

The question of cooptation is therefore not so clear-cut. Strategically, a symbiotic approach may be one way of realising radical social movement agency in the global political economy. The challenge for grassroots movements is to reradicalize NGOs and their message, which often becomes watered down through the process of institutionalization and engagement in global civil society, while maintaining vigilance against cooptation. Social movements are actively engaged in global environmental governance, sometimes shifting and influencing the agenda. In the case of the toxic waste trade, for example, social movement campaigning against the toxic waste trade contributed directly to the establishment of the 1989 Basel Convention. This institution itself, however, is an interstate institution with an end-of-pipe approach to "managing" the transboundary movement of toxic waste (Secretariat of the Basel Convention 1994, 8). It has not yet succeeded in collecting more than thirty-five out of sixty-two required ratifications for the ban to come into force (count as of January 2003). While the ban presents a small victory for environmental justice, it is in potential conflict with the WTO and has met with objections from industry, who are constantly attempting to undermine it (Housman et al. 1995, 146; Krueger 1999, 126).

Further, with the development of a ban on waste, 90 percent of all hazardous waste export schemes now involve a claim of "recycling" or "further use," (Puckett 1994, 56) a loophole that the Basel Convention attempted to tackle with a call to phase out the dumping and recycling of hazardous waste in poor countries by the end of 1997 (UNEP 1995, 16). This, however, is likely to lead to an increase in transfers of dirty technology and the movement of entire polluting industries to non-OECD countries. Already some evidence suggests that TNCs in the most hazardous industries (for example pesticides, asbestos, benzidine dyes, vinyl chloride, and lead smelting) are taking advantage of less stringent environmental regulations outside the OECD (Clapp 1994).

One of the ways the Basel Convention has attempted to further its mandate is through the establishment of Regional Centers for Waste

Management Training and Technology Transfer in the South as well as countries in Central and Eastern Europe (SBC 1994, 21). While this may enhance developing countries' ability to manage and dispose of their waste more efficiently, it remains as yet unclear what global role improved facilities in developing countries will play. In the context of the global political economy, it may increase the trade in hazardous wastes for recycling and disposal to developing countries and deter the need to deal with hazardous waste at source. Thus the agenda moves on, influenced by social movements, not always in expected ways, leading to new situations that require fresh strategies of resistance.

Conclusions

This chapter has sought to theorise the significance of global civil society as a global actor, while at the same time problematizing the view that this sphere is a homogeneous one. Global social movements are seeking to carve out a space among dominant corporate actors in order to challenge global environmental governance. They do so through a variety of strategies, some engaging with the global agenda in order to influence its direction, others taking a rejectionist stance against the totality of global capitalist hegemony. Rather than juxtaposing them as insiders versus outsiders, however, this chapter has argued that they may be seen as part of a complex ensemble of social forces. Furthermore, the chapter has argued that the agency of social movements involves more than merely participating in global environmental governance.

The chapter also argued that a conceptualization of "global" as causal rather than spatial is crucial in the realization of radical agency; global social movements, therefore, are those that adopt an analysis of global hegemonic formations and develop appropriate strategies. A collective global consciousness would bring the insiders and outsiders together in a symbiotic strategy. Some radical movements are pointing out the dangers of cooptation through global civil society, as well as the dangers of adopting orthodox discourse. They are recognizing the need for the engagers to retain links with the grassroots in the battle over the agency of global civil society and attempts to radicalize and expand it (Gorz 1987, 40). Similarly, grassroots movements that "think globally" while resisting locally, nationally, and transnationally across the globe can remain in dialogue with the institutionalized environmental organizations, keeping them alert and keeping the radical message of ecology and social justice alive by challenging the arbitrary limits of social reality and social practice.

Note

This chapter elaborates on an earlier article (Ford 2003).

References

Ashley, R. (1993). Imposing international purpose: Notes on the problematic of governance. In E. O. Czempial & J. N. Rosenau (Eds.), *Global changes and theoretical challenges* (pp. 251–290). Lexington, MA: D. C. Heath.

Bourdieu, P. (1977). *Outline of a theory of practice*. Cambridge, UK: Cambridge University Press.

Brown, P., & Ferguson, F. I. T. (1995). Making a big stink: Women's work, women's relationships, and toxic waste activism. *Gender and Society*, *9*(2), 145–172.

Brown, P., & Masterson-Allen, S. (1994). The toxic waste movement: A new type of activism. *Society and Natural Resources*, *7*(3), 269–287.

Clapp, J. (1994). The toxic waste trade with less-industrialised countries: Economic linkages and political alliances. *Third World Quarterly*, *15*(3), 505–518.

Clapp, J. (1998). The privatisation of global environmental governance: ISO 14000 and the developing world. *Global Governance*, *4*, 295–316.

Chatterjee, P., & Finger, M. (1994). *The earth brokers*. London: Routledge.

CGG (Commission for Global Governance) (1995). *Our global neighbourhood*. Oxford: Oxford University Press.

Cox, R. W. (1981). Social forces, states and world orders. *Millennium*, *10*(2), 126–151.

Cox, R. W. (1996). Gramsci, hegemony and international relations: An essay in method. In R. W. Cox & T. J. Sinclair (Eds.), *Approaches to world order* (pp. 124–143). Cambridge, UK: Cambridge University Press.

Doran, P. (1993). The Earth Summit (UNCED)—Ecology as spectacle. *Paradigms*, *7*(1), 55–65.

Eschle, C., & Stammers, N. (2001). Taking part: Social movements, INGOs and global change. Paper presented at the 5th Conference of the European Sociological Association, Helsinki.

Finger, M., & Tamiotti, L. (1999). New global regultory mechanisms and the environment: The emerging linkage between the WTO and the ISO. *IDS Bulletin, 30*(3), 8–15.

Ford, L. H. (1999). Social movements and the globalisation of environmental governance. *IDS Bulletin, 30*(3), 68–74.

Ford, L. H. (2003). Challenging global environmental governance: Social movement agency and global civil society. *Global Environmental Politics, 2,* 120–134.

Gibbs, L. M., & CCHW (The Citizens Clearinghouse for Hazardous Waste) (1995). *Dying from dioxin: A citizen's guide to reclaiming our health and rebuilding democracy.* Boston, MA: South End Press.

Gill, S. (1991). Reflections on global order and sociohistorical time. *Alternatives, 16*(3), 275–314.

Gorz, A. (1987). *Ecology as politics.* London: Pluto.

Gramsci, A. (1971). *Selections from the prison notebooks.* New York: International Publishers.

Grubb, M., Koch, M., Munson, A., Sullivan, F., & Thomson, K. (1993). *The "Earth Summit" agreements: A guide and assessment.* London: Earthscan.

Haas, P. (1990). Obtaining international environmental protection through epistemic consensus. *Millennium, 19*(3), 347–363.

Haas, P., Keohane R., & Levy, M. A. (Eds.) (1993). *Institutions for the Earth: Sources of effective international environmental protection.* Cambridge, MA: MIT Press.

Hallowes, D. (1993). *Hidden faces—Environment, development, justice: South Africa and the global context.* Scottsville: Earthlife Africa.

Housman, R., Goldberg, D., Van Dyke, B., & Zaelke, D. (Eds.) (1995). *The use of trade measures in select multilateral environmental agreements.* Geneva: UNEP.

ICC (International Chamber of Commerce) (1991). *The business charter for sustainable development: Principles for environmental management.* Paris: ICC.

Keck, M. E., & Sikkink, K. (1998). *Activists beyond borders: Advocacy networks in international politics.* Ithaca, NY: Cornell University Press.

Krasner, S. D. (Ed.) (1983). *International regimes.* Ithaca, NY: Cornell University Press.

Krauss, C. (1993). Women and toxic waste protests: Race, class and gender as resources of resistance. *Qualitative Sociology, 16*(3), 247–262.

Krueger, J. (1999) *International trade and the Basel Convention.* London: Earthscan/RIIA.

Kütting, G. (1997). Assessing the effectiveness of international environmental agreements: a critique of regime theory and new dimensions of analysis. Paper presented at the Annual Conference of the British International Studies Association, Leeds.

Lipschutz, R. D., with Mayer, J. (1996). *Global civil society and global environmental governance*. Albany, NY: SUNY Press.

Maclean, J. (1999a). Towards a political economy of agency in contemporary international relations. Draft paper.

Maclean, J. (1999b). Towards a political economy of agency in contemporary international relations. In M. Shaw (Ed.), *Politics and globalisation: Knowledge, ethics, agency*, (pp. 174–201). London: Routledge.

Paterson, M. (2001). *Understanding global environmental politics: Domination, accumulation, resistance*. London: Macmillan.

Porter, G., & Brown, J. W. (1991). *Global environmental politics*. Boulder, CO: Westview Press.

Puckett, J. (1994). *The Basel opportunity. Closing the last global waste dump*. Seattle: Greenpeace.

Puckett, J. (1999). *When trade is toxic: The WTO threat to public and planetary health*. Seattle: APEX/BAN. Available online at: <http://www.ban.org/>.

Saurin, J. (1993). Global environmental degradation, modernity and environmental knowledge. *Environmental Politics*, 2(4), 46–64.

Saurin, J. (1996). International relations, social ecology and the globalisation of environmental change. In J. Vogler & M. Imber (Eds.), *The environment and international relations* (pp. 77–98). London: Routledge.

SBC (Secretariat of the Basel Convention) (1994). *Basel Convention on the control of transboundary movements of hazardous wastes and their disposal, 1989 and decisions adopted by the first (1992) and the second (1994) meetings of the conference of the parties*. Geneva: UNEP/SBC/94/3.

Seager, J. (1993). *Earth follies: Feminism, politics and the environment*. London: Earthscan.

Szasz, A. (1994). *Ecopopulism—Toxic waste and the movement for environmental justice*. Minnesota: University of Minnesota Press.

Thomas, C. (Ed.) (1994). *Rio—Unraveling the consequences*. Ilford, UK: Frank Cass.

Trisoglio, A., & Kate, K. ten (1991). *From WICEM to WICEM II: A report to assess the progress in implementation of WICEM recommendations*. Geneva: UNEP.

United Nations (1992). *Agenda 21*. UNCED. Geneva: United Nations.

UNEP (1995). *Evaluation of the effectiveness of the Basel Convention*. Meeting of the Conference of the Parties of the Basel Convention, Geneva, UNEP/CHW.3/Inf.7.

Walker, R. B. J. (1994). Social Movements/World Politics. *Millennium: Journal of International Studies*, 23(3), 669–700.

Wapner, P. (1997). Governance in global civil society. In O. Young (Ed.), *Global governance: Drawing insight from the environmental experience* (pp. 65–84). Cambridge, MA: MIT Press.

Waterman, P. (2001). *Globalization, social movements and the new internationalisms*. London: Continuum.

Willetts, P. (Ed.) (1996). *The conscience of the world: The influence of NGOs in the UN system*. London: Hurst and Co.

Williams, M., & Ford, L. (1999). The World Trade Organisation, social movements and global environmental management. In C. Rootes (Ed.), *Environmental movements: Local, national, global* (pp. 268–289). London: Frank Cass.

Young, O. (1989). *International cooperation: Building regimes for natural resources and the environment*. Ithaca, NY: Cornell University Press.

Young, O. (Ed.) (1997). *Global governance: Drawing insights from the environmental experience*. Cambridge, MA: MIT Press.

Young, O. (1997a). Rights, rules and resources in world affairs. In O. Young (Ed.) *Global governance: Drawing insights from the environmental experience* (pp. 1–26). Ithaca, NY: Cornell University Press.

13

Business and International Environmental Governance: Conclusions and Implications

Peter J. Newell and David L. Levy

A political economy approach to understanding international environmental governance requires an examination of the interplay between industry economic structures, corporate strategies, and the development of systems of environmental governance. This volume has attempted to construct and elaborate such an approach, by exploring the role of business across a number of environmental issues, and by developing a conceptual framework for thinking about the complex, contested terrain of environmental governance. Here we draw out some of the key elements of a political economy approach, assess the relevance and adequacy of our theoretical perspective, and discuss some of the implications regarding future research agendas on environmental governance.

First and foremost, the contributors to this volume demonstrate that firms, in their roles as investors, innovators, experts, manufacturers, and polluters, are critical players in developing the architecture of global environmental governance. If we are serious about understanding the economic and political structures and processes that give rise to environmental problems, and simultaneously are expected to address them, we cannot afford to leave business out of our analysis. This seemingly obvious point does not resonate with predominant renditions of global environmental politics, which continue to employ a state-centric model of interstate bargaining in their analysis (see chapter 2). If we are to place firms centrally in our analysis, we need to look inside the firm to understand the processes by which they make decisions about technological choices and political positions. We also need to locate firms within the competitive dynamics of particular industry structures, and in the institutions and ideologies of the wider political economy.

The various cases in this volume amply demonstrate the ways in which the day-to-day production, research and marketing practices of firms are decisive in shaping the environmental impacts of human productive activity. A handful of large multinational corporations (MNCs) in the chemicals industry was responsible for the vast majority of production of ozone-depleting substances, as well as the development and commercialization of alternatives. The decisions of a few biotechnology companies determine which foods will be genetically modified, what procedures farmers wanting to plant seeds and apply pesticides have to follow, and which countries will have access to the technology. Private water companies, through their pricing and investment strategies, decide which communities have access to clean water and at what cost. The activities of these firms, however, do not just generate environmental problems to be regulated; a central argument of this book is that business activity is constitutive of environmental governance. Business is not just a subject of a regulatory system imposed by the state; rather, business is an intrinsic part of the fabric of environmental governance, as rule maker, and often rule enforcer. Dauvergne's study of the logging industry in Asia demonstrates how the governance system is actually dominated by informal corporate norms and logging practices on the ground, resulting in severe environmental impacts. If environmental problems are to be successfully addressed, it follows that the assent, if not active cooperation, of major sectors of industry is necessary. The change in the stance of a couple of large chemical companies helped to secure international agreement on the Montreal Protocol on Ozone Depleting Substances, and the successful implementation of this regime required business cooperation in developing alternative chemicals. In the climate change case, while some U.S.-based MNCs have successfully blocked U.S. ratification of the Kyoto protocol, it is notable that others are pursuing emission reduction technologies even in the absence of a binding regulatory framework.

Business does not play a passive role in this system of environmental governance, merely responding to market signals and governmental regulations. A key theme running throughout this volume is the agency of firms as political actors. Several of the chapters discuss the significant level of corporate political activity undertaken by large firms affected by environmental issues. Coen describes how political strategies have evolved in the face of changing industry structures and political contexts, in particular, the corporate trend toward Europe-wide operations and the increasing degree of regulation at the EU level. Industry associations active on environmental issues are increasingly issue rather than sector-based, and organized at the international rather than national level. Business political activity expresses itself in many forms. The chapters provide examples of conventional lobbying activity through formal and informal channels, financial donations to political campaigns, participation in advisory panels, and the provision of technical expertise. The importance of developing wider coalitions and alliances, with NGOs as well as other business sectors, is also discussed by several contributors. Partnering with NGOs and community groups is increasingly seen as key to the credibility of firm's social and environmental credentials.

A key contribution of this volume is to provide a broader understanding of power and the multitude of ways in which it is exercised by business actors. A number of contributors highlight the significance of discursive politics. Coen points to the efforts by business to gain access to policy forums by claiming to represent broad economic and societal interests rather than narrow sectoral concerns. This attempt to represent the general interest underlies strategies of building horizontal alliances and mobilizing "grassroots" groups, and it forms the basis of business demands for a special consultative mechanism at the UN. The quest for legitimacy is also one of the drivers of the development of private environmental management standards such as ISO 14000, described by Clapp.

Public debates concerning the science and economics of environmental issues have become key sites of political contestation over the risks, costs and benefits of different courses of action and inaction. Andrée demonstrates the ways in which biotech firms have sought to deflect fears raised by environmentalists about environmental and health impacts from GM crops by emphasizing the precision and predictability of the technology and by portraying their products as solutions to hunger, in an effort to shift debate from risks to potential benefits. Levy points to the considerable efforts by companies in fossil-fuel-related sectors to emphasize scientific uncertainties concerning climate change and to stress economic costs of litigation. Internal industry debates illustrate business awareness of the crucial role of discursive politics in securing legitimacy. While many companies were beginning to accept, if grudgingly, the link between greenhouse gas emissions and climate change, the American Petroleum Institute warned that abandoning the fight on climate science would entail a strategic concession of the central moral issue, and leave industry appearing to be selfishly concerned with its narrow economic interests. Language and knowledge are clearly key sites of contest in environmental politics.

Even business activities traditionally considered to be within the realm of markets have political dimensions. The decision by oil companies Shell and BP to invest in renewable energy technologies, discussed by Levy, not only complement the more accommodating stance of these companies in the climate regime negotiations, but also provide legitimacy as the companies seek influence on the emerging regulatory structure and a shield against public pressure for more stringent measures. Falkner challenges the idea that business passively adapts to the "technology-forcing" pressures of environmental regulation; rather, he argues persuasively that the innovative capacity of firms represents a form of technological power, which can play a critical role in shaping the design and phasing of environmental regulations. In areas where international regulatory structures are weak or absent, such as water and logging, environmental governance systems are dominated by the market activities of private firms, as they decide what products to sell in which markets and at what price. Ultimately, the traditional distinction between political and market strategies is unsustainable.

If firms are political actors performing these multiple functions, then the political process of negotiating and constructing systems of environmental governance needs to be revisited. Several of the contributions to this volume illustrate the complex processes of multiparty bargaining, alliance building, compromise, and accommodation that lead to the specific features of a given regime. Dauvergne's account of the logging industry describes how bargains and compromises with sociopolitical allies, such as politicians, bureaucrats, and military officers, as well as other corporate executives, have allowed loggers to dilute and deflect pressures for genuine environmental reforms. Both Levy and Andrée illustrate the rich interplay of moves and countermoves as firms and NGOs deploy a combination of economic, discursive, and organizational strategies. As corporate strategies become more sophisticated, so NGOs also learn to organize international alliances and partnerships, and to engage in science and policy debates.

The significance of opening up the negotiating process to scrutiny is that it reveals the contested and contingent nature of business influence, and suggests a degree of indeterminacy regarding outcomes. In the case of biotechnology, despite strong state backing, the industry has faced stringent regulation in Europe in the face of consumer concern, and a major multinational company was forced to back down, publicly at least, on the use of the controversial terminator technology that produces seeds capable of being used for one season only. In the climate case, industry efforts to question official scientific reports and to mobilize the appearance of grass roots citizen support backfired in the face of challenges from environmental NGOs. Falkner concludes that despite the pervasive influence of large chemical multinationals on the ozone regime, corporate actors were never in control; they only supported it as a second-best strategy, and sought to shape its evolution and implementation. In demonstrating their ability to develop new chemicals and meet lower production targets for CFCs, they inadvertently set the stage for states and NGOs to push for a comprehensive elimination of ozone-depleting substances. An important implication of this attention to the details of the bargaining process is a recognition that intelligent strategy is itself a source of power, one that has been neglected by scholars in international relations and organization theory.

One reason that the negotiating process itself is complex and indeterminate is that industry is not monolithic. Firms approach environmental issues with a conception of interests that is influenced by their location, the sector in which they are active, and their individual location within the competitive terrain. Differences between firms come to the fore when they are encouraged to formulate common policy positions as part of international environmental negotiating processes. Falkner's chapter on the ozone regime shows the important divergences in policy preferences that exist between firms at different stages of the value chain and with different relationships to technology. The story provides a strong case of business conflict, where there is evidence of heterogeneity even within producer and user groupings. Andrée's chapter on the biotechnology industry highlights a split between technology providers at the input end, such as Monsanto, and large grain traders such as Cargill. In the insurance industry, we see differences between insurers and reinsurers, as well as trans-Atlantic divisions.

Coen's chapter highlights some important differences between the lobbying styles, modes of organization, and institutional access of firms in Europe compared with their counterparts in the United States. Lobbying in the EU tends to be conducted more through private meetings and corridor conversations, whereas in the United States there are structured opportunities for more confrontational forms of engagement. In the case of climate change, the different responses of American and European firms in the same sector appear to be related to these different political structures and traditions of corporate political activity, though Levy also points to divergent expectations concerning the likely future of the regulatory environment and markets for low emission technologies. Given the many uncertainties concerning the future of regulations, technologies, and markets, some firms might find it wise and rational to invest substantially in renewable energy, while others find it equally rational to invest their resources in extending conventional technologies and in political efforts to block regulation. The broader implication is that business does not have a unified and objective set of interests; rather, corporate strategies are premised on perceptions of interests that are constructed within particular institutional environments.

The dynamic nature of the negotiation process, in which interests and alliances shift over time, is amply illustrated in the cases examined. Coen shows how business political activity in Europe adapts to the shifting competences and mandates of the EU institutions, and is increasingly influenced by the American style of aggressive lobbying. He describes how international quasi-policymaking organizations, such as the Trans-Atlantic Business Dialogue, are increasingly coordinating business positions on environmental issues. Levy also notes some convergence in trans-Atlantic corporate responses to climate change, and argues that new issue-specific industry associations and the institutions of the regime itself are responsible for instilling common perspectives on climate change among firms from diverse sectors and countries. The insurance industry on the other hand lost its initial interest in climate change mitigation with the development of new financial instruments to handle increased risks. These dynamics, combined with differences among firms, contribute to the instability of alliances. The Global Climate Coalition, for example, broke up after a number of key firms recognized that their legitimacy with policymakers and the public was threatened by continued membership of an organization that vehemently challenged the scientific basis for regulating greenhouse gases. More generally, it is the complexity and instability of environmental governance systems that create the space for actors to develop strategies that can leverage existing tensions and magnify their impact. This, of course, was the intention of Greenpeace's Jeremy Leggett as he tried to work with the financial sector against the fossil-fuel led industry alliance opposing controls on greenhouse gases.

Although the primary focus of this book has been on business as a political actor, it is clear that NGOs are also central players. Several contributors to the volume describe episodes in which environmental and other groups have played a key part in setting agendas, contesting the framing of issues and monitoring the implementation of an environmental regime. Using methods that are remarkably similar to those employed by business, NGOs pursue their goals by acquiring economic resources, building organizational capacity and alliances, developing formal and informal channels of access to policymakers, and by engaging in public debates concerning the science and economics of environmental issues. Where NGOs are lacking in material resources, they can sometimes compensate through their claims to social legitimacy and sophisticated strategy. Leggett's efforts to develop relationships between environmental NGOs and the insurance industry over the climate issue provide a good example. In another case, Greenpeace developed its own ozone-friendly "Greenfreeze" refrigerator to combat industry claims that CFC-free refrigerators were a technological impossibility. Rather than contesting policy through international fora, a direct challenge to the leading manufacturers of ozone-depleting chemicals put them on the defensive and forced a response. Of course, it is rare for civil society groups to achieve clear-cut successes. In the case of forests, very little progress has been made in constructing an effective global regime to reverse deforestation, aside from some weakly worded declarations. Dauvergne's chapter shows that close economic and social ties between policy and commercial elites at the national level are at least partly responsible for this intransigence. Ford shows how social movements have forged strategic connections with NGOs, working inside the Basel regime on toxic waste, often alongside developing country governments, amplifying their voice, and drawing attention to the local impacts of the global trade in toxics. Ultimately, though, many of their concerns have been screened out of the process through attempts to uphold technocratic approaches to decision making and concerted industry lobbying.

It would be a mistake, however, to portray international environmental politics as a business versus NGOs game. An important theme emerging from the chapters is the significance of horizontal issue-specific alliances that bring together business and NGOs. Through these relationships, environmental NGOs can offer business a measure of legitimacy, networks of contacts, and a degree of scientific expertise. Business can provide environmental groups, in return, with financial resources, global organizational reach, and the prospect of direct influence on the industrial practices that affect the environment. In the climate case, competition existed between two business-civil society alliances, one opposing mandatory controls and one supportive. The contribution by Ford most explicitly considers the role of NGOs, and argues that civil society should not be considered merely a counterweight to business, but rather a key site of political struggle.

A political economy approach demands that we pay attention not just to the environmental impacts of a particular governance system, but also to asymmetries and privilege during the negotiation process. Clapp's chapter shows how the decision-making process around ISO environmental management systems is dominated by the concerns of the better represented MNCs that attend the bodies' meetings, rather than those of the far more numerous, but less mobilized, small and medium enterprise sector. Equally, northern-based enterprises are much better represented than their southern counterparts. As a result, the ISO environmental standards are perhaps more significant as a barrier against competition from smaller companies and firms from the South, than as a framework for improving environmental performance. Similarly, Finger's chapter on the water industry describes how privatization and market pricing affects the availability of clean water in poor communities.

The development of new regimes has also shifted the balance of power among organizations. Insurance companies, by developing private financial mechanisms to manage risks, arrogate to themselves the decisions about who is to be made subject to large-scale climate related risks. Environmental groups willing to develop partnerships with business, such as World Resources Institute and Environmental Defense, have grown in resources and influence, though some might counter that they have given more, in terms of lending their credibility to business, than they have gained in the way of concrete change in business practices. The role of scientific advisory bodies such as the IPCC has certainly become more institutionalized and influential in global governance arrangements and, as a result, attracted the attention and increasingly involvement of firms in trying to shape the tone and presentation of the advice provided to policy-makers.

Finally, a political economy approach requires us to locate regimes in the global economic and political system of which they are a part. The contributions in this book demonstrate the prevalence, across dissimilar issue areas, of common patterns and practices, such as the creation of new property rights and trading mechanisms, new forms of business-NGO-state partnerships, and the development of private codes and standards. Above all, business has gained ideological legitimacy as a responsible steward of the environment, willing to deploy financial resources and technological expertise to address environmental issues. Yet the relationship between macrostructures and individual regimes also signals the limitations of pluralist approaches that view regime negotiations as bargains amongst equals within isolated arenas. Businesses are clearly not always able to secure outcomes favorable to their interests, but they do appear to enjoy a privileged position that draws from the power of international institutions governing trade and finance, a pervasive discourse supportive of markets and deregulation, and the support of states committed to "business competitiveness."

Andrée's chapter illustrates the key role of trade regimes in shaping the ways in which the Cartagena Protocol on Biosafety governs the transboundary transfer of GMOs. The support leant to the major manufacturers of GMOs by their host countries in North American and Europe is a function of the central role of biotechnology in the knowledge-based economies that those countries are seeking to develop. Dauvergne describes how the management of tropical forests is intertwined with the political and economic structures of many SouthEast Asian countries, particularly the ties between commercial logging interests and political elites. Governments are extremely reluctant to enforce regulations against powerful timber producers whose revenues underpin their country's economy. Individual regimes are also constitutive of international political economic structures; for example, the efforts of the biotechnology industry to strengthen intellectual property rights for genetically modified organisms (GMOs) have helped to institutionalize the rights afforded to firms under the Trade Related Intellectual Property Rights (TRIPs) agreement of the WTO.

The State in International Environmental Governance

The shifting nature of relations among business, the state, and civil society is illuminated by the studies in this volume. One important theme is the ongoing privatization of environmental governance. This is evident in the unprecedented influence of the fossil fuel lobby over U.S. climate policy, in the technological power of chemical companies in shaping the ozone regimes, and in the acquiescence of regulatory agencies such as the U.S. Food and Drug Administration to the agenda of the biotechnology industry. It is also evident in the growth of private codes and standards, such as ISO 14000 described by Clapp, and in the quasi-private policy-setting role of bodies such as the Trans-Atlantic Business Dialogue, discussed by Coen. Finally, as Finger's chapter on the water industry makes clear, privatization transfers environmental governance functions to the corporate sector. Yet, most of the contributors concur that this process of privatization does not necessarily signify a decline in the authority or regulatory capacity of the state; rather, the state's relationships with civil society and the corporate sector is being redefined in complex ways. Finger describes how water multinationals actively leverage the considerable resources and regulatory authority of states in order to obtain investment, revenues, social and economic stability, and an appropriate regulatory environment.

While many issue areas reflect the general trend toward deregulation and the outsourcing by states of monitoring and reporting functions to private actors, several authors also note a simultaneous increase in state regulatory capacity. The privatization of water and electricity, for example, require a whole slew of regulations to address competition, pricing, and security of supply in these quasi-monopolistic industries. The establishment and enforcement of new forms of intellectual property rights, and the development of market mechanisms such as emission trading schemes, also require a complex administrative structure. At the same time, micromanaging the global trade in GM seed and establishing effective biosafety regulations, as required by the Cartagena Protocol, stretches the regulatory capacity of many governments.

Because of their reliance on expertise, resources, and capacity support from global actors, many state-based environment agencies are globalizing in new ways. State agencies are nested, to different degrees, within international networks of environmental bureaucrats, economists, and scientists based in a range of different organizations. In the EU case, Coen shows how the environment Directorate Generale of the European Commission (DG-XI) seeks to buffer its weak position within the policy process though coalition-building with environmental NGOs. This blurring of boundaries is accompanied by the diffusion of ideologies through institutions such as the World Bank and Business Council for Sustainable Development that discredit "command-and-control" forms of regulation as overly intrusive, while portraying businesses as socially responsible corporate citizens who are entitled to a role as partners in environmental governance, not just its subjects. Overall, this volume indicates that the state is, if anything, expanding its overall regulatory scope, often with the help of international institutions in the case of developing countries, but in ways that are penetrated by expanding market relations and their associated ideologies, and increasingly enmeshed in a web of relationships with firms, NGOs, and international organizations in systems of environmental governance.

Evaluating the Neo-Gramscian Framework

We proposed in chapter 2 that a neo-Gramscian theoretical framework could be of value in bringing coherence and insight to understanding the role of business in international environmental governance. Levy's chapter on climate change and Andrée's on biotechnology make the most detailed and explicit use of this framework, though Jagers, Paterson and Stripple, Clapp, and Ford each deploy elements of a Gramscian analysis in their contributions. While other chapters employ a more general political economy perspective, Falkner adopts perhaps the most pluralist viewpoint in this volume, and emphasizes the indeterminacy of outcomes in his discussion of the technological power of business in the ozone case. If the political economy of environmental governance can be examined without explicit mention of Gramscian concepts, it is legitimate to question how much our framework adds to the analysis. One does not need to refer to hegemony and historical blocs to reach most of the key insights of this volume, such as the pervasive influence of business, the legitimating role of civil society, or the significance of business strategies.

By themselves, however, these insights have little more status than a summary of empirical observations drawn from a range of case studies. We suggest that the neo-Gramscian framework integrates these insights in a more theoretically grounded and intellectually satisfying manner. It provides a more systemic understanding of dynamic processes of political contestation over environmental governance, and their linkages to macropolitical and economic structures. This understanding clarifies why, across a number of issue arenas, similar patterns of political action and accommodation can be observed. The concept of an historical bloc, for example, indicates that the construction of a dominant coalition in a particular issue arena is not just a matter of assembling a coalition of groups, but involves particular economic arrangements, shifting the framing of actors' interests, and the development of new organizational forms. The conceptual framework therefore helps elucidate the common patterns of economic, discursive, and organizational strategies that we see firms and NGOs deploying across multiple issues, from climate change to genetic engineering.

In some of the cases in this volume, the neo-Gramscian approach helps to account for phenomena that would otherwise be difficult to explain. In the climate change case, the corporate shift toward a more accommodating position cannot be attributed to major breakthroughs in low-emission technologies, new scientific discoveries about the seriousness of global warming, or the inevitability of strong regulation. Instead, Levy makes the case that a series of relatively small changes in the economic, discursive, and organizational spheres made it impossible for industry to construct a hegemonic position that would align the interests of major industrial and social groups around opposition to mandatory controls. Another example is the failure of corporate "astroturf" strategies, which mobilize and represent "artificial" grassroots opinion; while these efforts might provide some short-term cover for politicians, they also indicate the failure of corporate public relations efforts to gain authentic legitimacy in civil society. At the same time, in almost every issue area, companies have succeeded in maintaining their market position, and even increased their social legitimacy and political influence, while making a series of concessions to new regulatory structures.

The neo-Gramscian approach holds the promise then of bringing theoretical richness and sophistication to the study of environmental governance. Conventional approaches tend to be stymied somewhat by stagnant debates over structure versus agency, elite domination versus pluralism, and objectivism versus constructivism. The Gramscian concept of hegemony goes some way toward resolving these dualisms, in its depiction of the tensions between business dominance and the contested, contingent nature of that dominance, between structural sources of influence and the potential for agents to change those structures, and between economic roots of power and the ideological formations that legitimize the exercise of that power. The risk here is that it might appear to be all things to all people rather than a meaningful synthesis. We suggest, however, that the theory opens up a more detailed understanding of these issues, so that we can begin to comprehend the conditions that allow business to "win" and those that lead to compromise, which governance structures are stable and which are ripe for change, and which strategies are likely to succeed and which to fail.

Whether we invoke Gramsci or not, a political economy approach brings a fresh and innovative perspective to understanding systems of environmental governance, one that is likely to stimulate new questions and a search for new connections. Most importantly, perhaps, it enables us to conceptualize the development of systems of environmental governance not as a rational problem-solving activity, but as a political effort to coalesce an alliance of groups around a specific set of arrangements. Governance, then, is not just imposed by states as a regulatory structure overlaying the affairs of consumers and firms; rather, governance arises out of the interactions among business, NGOs, state agencies, and international organizations. This leads to our broader conception of governance as the ensemble of practices and processes, embedded within institutions and norms, which provide structure, pattern and order to activities with environmental impacts. A privatized, market-based water industry, for example, thus also represents a regime of governance. This conception allows us to analyze the political economy of different governance mechanisms, and ask questions about their distributional and environmental impacts. Moreover, a conception of governance as a complex, political-economic system enables us to examine its stability, and to probe the tensions and fissures that can give rise to instability and change. Finally, by understanding these instabilities and key points of leverage, the strategic dimension of power becomes significant in advancing or contesting particular forms of governance.

What this suggests is that future research agendas on global environmental politics would do well to develop the theorization, as well as empirical documentation, of the reciprocal links between corporate strategy and regimes of environmental governance. We have suggested that insights from organizational theory and corporate strategy might be fruitfully combined with analytical tools from International Political Economy to forge and understand these connections. As well as looking within the firm, this implies the need to relate particular practices of power within a regime of environmental governance to the broader structures of political and economic power in which they are embedded. Studies of issue-specific regimes, at whatever level, that locate their origins and development within a historical perspective and seek to account for the material relations that give rise to the regime, are likely to yield important insights into who is exercising power, the strategies by which it is exercised, and the environmental and distributional implications. The insights from this volume also suggest the need for a systemic analytical point of departure that examines conflict and coordination across networks and alliances, rather than the more familiar levels of analysis that are said to separate domestic from international political arenas, state from nonstate, and public from private. Perhaps especially in a context of globalization these distinctions become increasingly fragile.

We hope that that this book has raised a set of questions and suggested some tools that provide multiple entry points into a more textured and critical debate than we have had to date about the role of business in global governance, a debate that extends beyond the arena of environmental politics. It has certainly not attempted to bring closure to this discussion; rather our aim has been to sketch the contours of what we believe to be a key area of research informed by a series of rich empirical case studies. We are confident that this is a research agenda that students of global politics cannot afford to ignore, because it taps into a more fundamental set of questions about how we want to be governed globally and by whom. The work in this volume demonstrates vividly that the business of business is increasingly global governance and that governing globally increasingly demands the involvement and support of business.

Index